Bioenergetics

The Practical Approach Series

SERIES EDITORS

D. RICKWOOD
Department of Biology, University of Essex
Wivenhoe Park, Colchester, Essex CO4 3SQ, UK

B. D. HAMES
Department of Biochemistry and Molecular Biology
University of Leeds, Leeds LS2 9JT, UK

Affinity Chromatography
Anaerobic Microbiology
Animal Cell Culture
 (2nd Edition)
Animal Virus Pathogenesis
Antibodies I and II
Basic Cell Culture
Behavioural Neuroscience
Biochemical Toxicology
Bioenergetics
Biological Data Analysis
Biological Membranes
Biomechanics — Materials
Biomechanics — Structures and
 Systems
Biosensors
Carbohydrate Analysis
 (2nd Edition)
Cell–Cell Interactions
The Cell Cycle
Cell Growth and Division
Cellular Calcium
Cellular Interactions in
 Development

Cellular Neurobiology
Centrifugation (2nd Edition)
Clinical Immunology
Computers in Microbiology
Crystallization of Nucleic Acids
 and Proteins
Cytokines
The Cytoskeleton
Diagnostic Molecular Pathology
 I and II
Directed Mutagenesis
DNA Cloning: Core Techniques
DNA Cloning: Expression
 Systems
Drosophila
Electron Microscopy in Biology
Electron Microscopy in
 Molecular Biology
Electrophysiology
Enzyme Assays
Essential Developmental
 Biology
Essential Molecular Biology I
 and II

Bioenergetics
A Practical Approach

Edited by

GUY C. BROWN

*Department of Biochemistry,
University of Cambridge,
Tennis Court Road,
Cambridge CB2 1QW*

and

CHRIS E. COOPER

*Department of Paediatrics,
University College London Medical School,
The Kayne Institute,
5 University Street,
London WC1E 6JJ*

OXFORD UNIVERSITY PRESS
Oxford New York Tokyo

Oxford University Press, Walton Street, Oxford OX2 6DP

Oxford   New York
Athens   Auckland   Bangkok   Bombay
Calcutta   Cape Town   Dar es Salaam   Delhi
Florence   Hong Kong   Istanbul   Karachi
Kuala Lumpur   Madras   Madrid   Melbourne
Mexico City   Nairobi   Paris   Singapore
Taipei   Tokyo   Toronto
and associated companies in
Berlin   Ibadan

Oxford is a trade mark of Oxford University Press

Published in the United States
by Oxford University Press Inc., New York

A catalogue record for this book is available from the British Library

Library of Congress Cataloging in Publication Data
Bioenergenetics : a practical approach / edited by Guy C. Brown and
Chris E. Cooper.
(The Practical approach series)
Includes bibliographical references.
1. Bioenergenetics—Methodology.   2. Biological transport, Active—
Research—Methodology.   I. Brown, Guy.   II. Cooper, Chris.
III. Series.
QH510.B53 1995   574.19'121–dc20   94–46232
ISBN 0 19 963489 0 (Hbk)
ISBN 0 19 963488 2 (Pbk)

Typeset by Footnote Graphics, Warminster, Wilts
Printed in Great Britain by Information Press Ltd, Eynsham, Oxon.

Preface

Bioenergetic research is at a crossroads. With the general acceptance of Mitchell's chemiosmotic hypothesis in the early 1980s the nature of the questions bioenergeticists have been required to answer has changed. Much research has focused on the search for the molecular mechanisms of ion-translocating proteins. On the other hand, the advent of new technologies has allowed us to probe bioenergetic processes in ever more physiological conditions. These two branches of bioenergetics, molecular mechanisms and physiological bioenergetics, are both at the forefront of the legacy left us by Peter Mitchell.

At the same time, the maturation of the field of bioenergetics has led to cross-fertilization with other fields. And this has led to both the necessity for scientists in other fields (e.g. medical scientists) to learn the classical bioenergetic techniques, and the application of foreign techniques to bioenergetics (e.g. molecular biology, electrophysiology, NMR).

One of the main purposes of this book is to describe the core of bioenergetic techniques that are necessary in analysing membrane translocating events. These include the measurement of the main fluxes and stoichiometries (Chapter 1), membrane transport (Chapter 2), protonmotive force (Chapter 3), and redox states and potentials (Chapter 5). The isolation of bioenergetic preparations is described for mitochondria and mitochondrial subcomponents (Chapters 7 and 8), and chloroplast subcomponents (Chapter 9), as well as the reconstitution and characterization of proteins in proteoliposomes (Chapter 4). Many of these techniques are old, but have required continual refinement to overcome pitfalls. It is hoped that these techniques will assist a new generation of experimenters, whether studying novel bioenergetic systems or perturbations of well-understood ones.

However, bioenergetics is not a science rooted in the past. New techniques are pushing back the boundaries of the subject. The mitochondrial membranes can now be patch-clamped to reveal single molecular events (Chapter 7). NMR can be used to measure levels, fluxes, and stoichiometries of bioenergetic intermediates *in vivo* (Chapter 8). EPR can be used to look within molecules (Chapters 5 and 9). Metabolic control analysis and related methodologies can be used to analyse the relation between changes at the molecular (or sub-molecular) level and behaviour of bioenergetic systems at the physiological level (Chapter 6). Chapter 9 illustrates how combining molecular biological and biophysical techniques is a powerful tool in dissecting the molecular mechanisms of photosynthetic reaction centres.

Because bioenergetics is such a broad subject covering such a wide-range of systems, we can not hope to cover anything like the full range of techniques

Preface

used. A number of other excellent books in the Practical approach series cover particular bioenergetic systems or techniques (e.g. Mitochondria, Photosynthesis: energy transduction, Biological membranes, Spectrophotometry and spectrofluorimetry). We therefore prefer to concentrate on the core techniques of bioenergetics, plus some more recent methods not covered in other books. Although molecular biology has made a huge contribution to bioenergetics recently, the methods involved are well described in other books in the series.

London   G. C. BROWN
August 1994   C. E. COOPER

Contents

Contents

9. Isolation and characterization of
photosynthetic reaction centres from
eukaryotic organisms   183

*Michael C. W. Evans, Beverly J. Hallahan, Jonathan A.
Hanley, Peter Heathcote, Nicola J. Gumpel, and Saul Purton*

Contents

Contributors

CRISTINA BALLARIN
Dipartimento di Chimica Biologica, Università di Padova, Via Trieste 75, 35121 Padova, Italy.

MARTIN D. BRAND
Department of Biochemistry, University of Cambridge, Tennis Court Road, Cambridge CB2 1QW, UK.

KEVIN M. BRINDLE
Department of Biochemistry, University of Cambridge, Tennis Court Road, Cambridge CB2 1QW, UK.

GUY C. BROWN
Department of Biochemistry, University of Cambridge, Tennis Court Road, Cambridge CB2 1QW, UK.

RICHARD CAMMACK
Metals in Biology and Medicine Centre, Division of Life Sciences, King's College London, Campden Hill Road, London W8 7AH, UK.

MICHAEL C. W. EVANS
Department of Biology, University College London, Gower Street, London WC1E 6BT, UK.

ALEXANDRA M. FULTON
Department of Biochemistry, University of Cambridge, Tennis Court Road, Cambridge CB2 1QW, UK.

NICOLA J. GUMPEL
Department of Biology, University College London, Gower Street, London WC1E 6BT, UK.

BEVERLY J. HALLAHAN
Department of Biology, University College London, Gower Street, London WC1E 6BT, UK.

JONATHAN A. HANLEY
Department of Biology, University College London, Gower Street, London WC1E 6BT, UK.

PETER HEATHCOTE
School of Biological Sciences, Queen Mary and Westfield College, Mile End Road, London E1 4NS, UK.

Contributors

PETER C. HINKLE
Section of Biochemistry, Molecular and Cellular Biology, Cornell University, Ithaca, NY 14853, USA.

EMMALEE V. MARSHALL
Department of Microbiology and Guelph-Waterloo Centre for Graduate Work in Chemistry, University of Guelph, Guelph, Ontario N1G 2W1, Canada.

SAUL PURTON
Department of Biology, University College London, Gower Street, London WC1E 6BT, UK.

M. CATIA SORGATO
Dipartimento di Chimica Biologica, Università di Padova, Via Trieste 75, 35121 Padova, Italy.

RALF T. VOEGELE
Department of Microbiology, University of Guelph, Guelph, Ontario, N1G 2W1, Canada.

SIMON-PETER WILLIAMS
Department of Biochemistry, University of Cambridge, Tennis Court Road, Cambridge CB2 1QW, UK.

JANET M. WOOD
Department of Microbiology, University of Guelph, Guelph, Ontario N1G 2W1, Canada.

JOHN M. WRIGGLESWORTH
Metals in Biology and Medicine Centre, Division of Life Sciences, King's College London, Campden Hill Road, London W8 7AH, UK.

Abbreviations

ADP/O ratio	moles of ADP consumed per mole of oxygen atoms reduced to water
ANT	adenine nucleotide translocator
asolectin	soybean phospholipid
BSA	bovine serum albumin
BTPP	butyltriphenylphosphonium
C	capacitance
C_1^i	control coefficient of i over j
CMC	critical micelle concentration
ΔG	Gibbs energy change at specified reactant and product concentrations
ΔH_{pp}	linewidth of EPR signal (peak to peak distance of 1st derivative spectrum)
$\Delta\mu_{H_+}$	proton electrochemical potential across a membrane
Δp	protonmotive force
ΔpH	pH difference between two phases separated by a membrane
$\Delta\psi$	membrane potential (electrical potential difference across a membrane)
$\Delta\psi_m$	membrane potential across the mitochondrial membrane
$\Delta\psi_p$	membrane potential across the plasma membrane
$\Delta\mu_s$	solute electrochemical potential gradient across a membrane
DCIP	2,6-dichloroindophenol
DCMU	3-(3,4-dichlorophenyl)-1,1-dimethylurea
DCPIP	2,6-dichlorophenolindophenol
$diSC_3$-5	3′,3′-dipropylthiodicarbocyanine
DMBQ	2,3-dimethylbenzoquinone
DPMIB	2,5-dibromo-3-methyl-6-isopropyl-1,4-benzo-quinone
DTT	dithiothreitol
E	redox potential at any specified set of component concentrations
$E^{0'}$	redox potential with components in their standard states (1M solution concentrations at 1 atm gases) except that pH = 7

Abbreviations

EPR	electron paramagnetic resonance
F	Faraday constant (96.5 kj V^{-1})
FCCP	carbonyl cyanide p-trifluoromethoxyphenyl-hydrazone
GAPDH	glyceraldehyde-3-phosphate dehydrogenase
HTG	n-heptylthioglucoside
H^+/ATP ratio	moles of protons translocated per mole of ATP formed
H^+/O ratio	moles of protons translocated per mole of oxygen atoms reduced
I_{max}	Inhibitor concentration just sufficient to cause maximum inhibition
LHCP	light harvesting complex
LHCPII	light harvesting complex of photosystem II
MDP	methylenediphosphonate
M_z	magnetization along the z axis
M_o	magnetization along the z axis at thermal equilibrium
NMR	nuclear magnetic resonance
OGP	octylglucopyranoside
PCr	phosphocreatine
PEP	phosphoenolpyruvate
PFK	phosphofructokinase
PGK	phosphoglycerate kinase
Ph^-	reduced phaeophytin
P_i	inorganic phosphate
P/O ratio	moles of ATP formed per mole of oxygen atoms reduced to water
PPBQ	phenyl-1,4-benzoquinone
PSI	photosystem I
PSII	photosystem II
P680	reaction centre chlorophyll in photosystem II
P700	reaction centre chlorophyll in photosystem I
Pyranine	8-hydroxy-1,3,6-pyrene trisulfonate
R	The gas constant (8.314 J mol^{-1} K^{-1})
RCR	respiratory control ratio (state 3 rate respiration rate divided by state 4 rate)
state 3	respiration rate following addition of ADP
state 4	resting respiration rate following conversion of ADP to ATP
T	temperature (Kelvin)
T_1	spin-lattice relaxation time
T_2	spin-spin relaxation time
TCA	trichloroacetic acid

Abbreviations

TMA	tetramethylammonium
TMAOH	tetramethylammonium hyroxide
TMPD	N,N,N′,N′-tetramethy-*p*-phenylenediamine
TPP	tetraphenylphosphonium
TPMP	triphenylmethyphosphonium

1

Oxygen, proton and phosphate fluxes, and stoichiometries

PETER C. HINKLE

1. Introduction

The measurement of coupling ratios should never be considered a routine undertaking, for history shows that practically every example went through several stages of eliminating systematic errors before the current relative consensus was reached. For example, traditionally the P/O ratio of mitochondria with succinate was thought to 'approach' 2.0, but currently it is considered to be 1.5 (1). This is partly due to the fact that measurements of oxygen and protons are rather prone to error. Errors may also result from the frame of mind of the experimenter. Since ratios are dimensionless and reflect the mechanism of the coupling process they are hard to view with objective detachment. I will not review the sad history of errors in ratio determinations, which have happened to most workers in the field, including myself, but will describe examples of what seem to be the most reliable methods currently used to measure coupling ratios and some of the sources of error which should be avoided.

2. P/O ratios

This oldest ratio, the moles of ATP formed per mole of oxygen atoms reduced to water, was originally measured with tissue slices or homogenates in a Warburg manometer (2). This instrument was difficult to use but had the advantage that long reaction times were possible and in some situations might still be the best method. Most measurements of oxygen uptake are now made with the oxygen electrode.

2.1 Oxygen electrode

The Clarke oxygen electrode is available commercially (e.g. Yellow Springs Instrument Co., Rank Brothers). To function properly care must be taken in the design and use of the cell that it is mounted in as well as with the electrode itself. The electrode consists of platinum and silver chloride surfaces covered with a drop of 2 M KCl and covered with a tight fitting thin Teflon membrane. The platinum electrode is polarized at -0.6 to 0.7 V relative to the silver chloride electrode, and the current flowing to maintain the polarization is proportional to the oxygen diffusion through the Teflon membrane. Membranes of various thickness are available, with thin membranes giving a faster response time and a higher noise level. The response time is generally quite slow, on the order of three to ten seconds to establish a new rate. The response and noise level are also a function of the efficiency and constancy of stirring of the solution, respectively. High noise is usually caused by poor placement of the magnetic stirrer which causes the stirring bar to occasionally hit the wall and stop. Visual inspection of the apparatus during measurements is important to detect problems with the stirrer or bubbles which must be avoided.

The electrode is best used in a glass thermostatted chamber of 1–3 ml volume (e.g. Gilson) as shown in *Figure 1*, but Plexiglass (Perspex) chambers are acceptable. Plastic chambers are more difficult to clean properly, since hydrophobic uncouplers and inhibitors are often used which are best washed from the chamber with 95% ethanol which causes cracks in plastic. Chambers with any Teflon parts touching the measured solution, which were once common, are not acceptable. Teflon contains a large amount of dissolved oxygen, which can diffuse into the solution being measured. The high solubility of oxygen in Teflon is why Teflon membranes are used in the electrode, but other Teflon surfaces, such as a stirring bar, must be avoided.

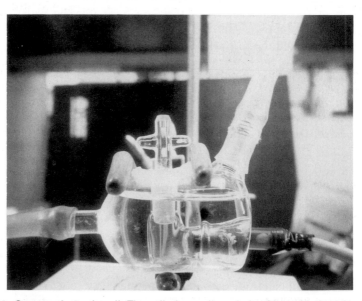

Figure 1. Oxygen electrode cell. The cell glass cell made by Gilson Medical Electronics. This cell together with the Yellow Springs electrode have been used by most American biochemists since about 1970.

The Teflon stirring bar is sometimes a major cause of error, especially if calibration based on equilibrium with air is used. Medium sized 'micro' bars 10×2 mm contain 65 nmol of oxygen, which equilibrates with the medium with a half-time of 70 sec (1). This can be a significant (10%) fraction of the total oxygen in the cell, which is available to drive phosphorylation but is not included in the calculation, causing an overestimation of the correct value of P/O. To avoid this problem glass covered stirring bars can be made. A short section of a thick paper-clip (e.g. 1 mm $\times$ 5 mm), or other wire found to be well attracted to a magnet, is dropped down a Pasteur pipette so that it becomes lodged in the thin tip. It can then be pulled out in a Bunsen burner flame to make a smooth sealed bar, with the glass cover tightly fitting the iron bar. If the glass is larger than the bar the magnet will not be as effective at driving the bar.

2.2 Calibration of the oxygen electrode

It is important to realize that the oxygen electrode measures activity of oxygen, not concentration. All solutions equilibrated with air at the temperature of the electrode cell (e.g. 25°C) will give the same reading on the electrode even though the concentration could be quite different, because solutes generally decrease the solubility of oxygen in an aqueous solution. For example, oxygen is about five times more soluble in ethanol than in water and yet ethanol equilibrated with air at 25°C gives the same reading as water equilibrated with air at 25°C. On the other hand, when an inhibitor such as rotenone, which is always dissolved in ethanol, is added to an aqueous solution (3 μl added to 2 ml, for example) the oxygen concentration is seen to rise because the extra oxygen dissolved in the ethanol is now seen by the electrode in water where the activity is higher. The same effect occurs when ice-cold aqueous solutions are added to a reaction medium at 25°C, because the solubility of oxygen in water at 0°C is about two times that at 25°C.

With the above principles established, it can now be seen that several methods can be used to calibrate an oxygen electrode, but that each calibration will apply only to one particular medium and temperature. Described below is a protocol for calibrations based on previously determined values for oxygen solubility (*Protocol 1*). However, if an unusual medium is used the oxygen content will have to be determined, for example by the oxidation of a known concentration of NADH (*Protocol 2*).

Protocol 1. Calibration of oxygen electrode based on previously
determined values for solubility of oxygen

1. Choose a medium and temperature used by previous workers. [a]

2. Correct the value for oxygen concentration for barometric pressure
   (BP) by multiplying times BP/760. [b]

Protocol 1. *Continued*

3. Equilibrate the medium with air by bubbling or shaking while the temperature is maintained by a water-bath for at least 30 min.[c]

4. Set the zero point on the recorder by filling the cell with medium, adding a few grams of sodium dithionite ($Na_2S_2O_4$), and when the electrode has responded completely adjusting the point on the chart to zero.

5. Rinse out the $Na_2S_2O_4$ with water, add air-equilibrated medium, and set the sensitivity knob on the electrode to about 95% full-scale on the recorder.

[a] For example, 250 mM sucrose, 8 mM KP_i pH 7.2, 4 mM $MgCl_2$, 0.08 mM EDTA, 0.15 mg/ml bovine serum albumin, contains 480 μM oxygen (3) (O, not O_2), 150 mM KCl contains 498 μM O (4), and 60 mM KCl + 125 mM sucrose contains 475 μM O (5) at 25°C and 760 mm Hg barometric pressure (see also refs 6 and 7).
[b] Remember that barometric pressure depends on altitude and that values reported by weather bureaus are corrected to sea-level unless requested otherwise.
[c] If the medium has been stored in a refrigerator, the equilibration at higher temperature will decrease the oxygen concentration. Ensure that the temperature bath for the electrode cell is set to the same temperature.

Protocol 2. Oxygen electrode calibration based on NADH oxidation

1. Calibrate an NADH solution (at pH > 8) in a spectrophotometer at 340 nm following oxidation (ΔE = 6.22 mM/cm) with alcohol dehydrogenase and acetaldehyde as described by Lemasters (8).[a]

2. Place the air-equilibrated solution in the oxygen electrode and add a solution capable of completely oxidizing 0.1 mM NADH in less than 1 min.[b]

3. Add aliquots of calibrated NADH sufficient to cause about 20% of the oxygen to be reduced (i.e. about 100 μM), and calculate the electrode sensitivity from the amount of NADH added and the deflection of the recorder pen.[c]

[a] The oxidation is necessary because some preparations of NADH contain UV absorbing material which is not NADH.
[b] If animal mitochondria or bacteria are being used, a preparation of everted vesicles (SMP) can be made by sonication. If plant mitochondria are used they will oxidize NADH without further treatment. Alternatively a solution of 10 μg/ml phenazine methosulfate and 0.5 μU/ml catalase can be used.
[c] During this calibration the linearity of the electrode and possible back diffusion of oxygen from the atmosphere or plastic in the electrode apparatus can also be checked by giving multiple additions and watching whether the trace drifts upwards after the NADH is oxidized.

2.3 Measurement of ADP/O ratios

With preparations which show respiratory control, such as rat liver mitochondria but including submitochondrial particles (9), the simplest method
to measure the stoichiometry of oxidative phosphorylation is to add a known
ADP pulse of about 400 nmol made with a precision microsyringe (e.g. 10
or 25 μl, Hamilton), and measure the resultant oxygen uptake (6, 10). In
such experiments the ADP solution should be assayed for contaminating
AMP (1) and the effective ADP concentrations taken as ADP + (2 × AMP),
since in the presence of Mg^{2+} ions and adenylate kinase activity, which is
usually present, AMP will be phosphorylated twice to form ATP. The total
concentration of adenine is best measured by the absorbance at 259 nm
(E mM = 15.4), and the correction for AMP applied from the per cent
contamination determined by enzymatic (11) or HPLC (1) methods. When
an ADP pulse is added to mitochondria respiring in a medium containing substrate and P_i, a burst of rapid oxygen uptake is seen as shown in *Figure 2*.

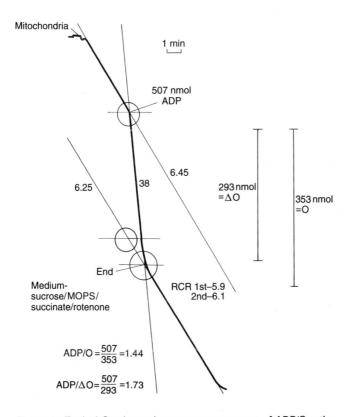

Figure 2. Typical O_2 electrode trace measurement of ADP/O ratio.

This actual trace from an oxygen electrode shows the typical signal-to-noise ratio. For maximum accuracy a large ADP pulse should be given, and the ADP concentration in the stock solution should be high (100 mM). This minimizes error from the fact that there is oxygen in the ADP solution which is usually kept at ice temperature, causing the oxygen concentration to be twice that at room temperature. If a 10 mM solution of ADP were used for the ADP addition an underestimate of the P/O ratio by 5% to 10% would be caused by the extra oxygen in the ADP solution, depending on the concentration of oxygen in the cell when the ADP pulse was given.

The end-point of the oxygen burst is traditionally determined by the extrapolation shown in *Figure 2*. This method to determine the oxygen used to phosphorylate the added ADP, called total oxygen, is actually a small underestimate because the state 4 rate is extrapolated back to the intersection with the state 3 rate, even though all of the ADP was not phosphorylated at the end-point thus determined. Some have used a more extensive extrapolation of the state 4 rate back to the time of ADP addition called ADP/ΔO. This method gives an overestimation of the true P/O ratio because the proton leak through the membrane driven by the electrochemical proton gradient Δp, is much less in state 3, when Δp is lower, than in state 4 when Δp is high. The dependence of the proton leak on Δp shows a non-ohmic proton conductance (12), which explains many of the complex relationships between flows and forces in mitochondria, and the fact that mitochondria with poor respiratory control still phosphorylate ADP with almost normal P/O ratios. *Protocol 3* takes the above considerations into account in measuring ADP/O ratios.

Protocol 3. Calculation of ADP/O ratios

1. Add air-equilibrated medium (250 mM sucrose, 8 mM KP$_i$ pH 7.2, 4 mM MgCl$_2$, 0.08 mM EDTA, 0.15 mg/ml bovine serum albumin) to the oxygen electrode cell.

2. Add mitochondria to give about 1 mg/ml final concentration.

3. Add substrate (e.g. 5 mM succinate) and inhibitors (e.g. 1 μM rotenone in a few microlitres of ethanol). Measure the rate of oxygen uptake for about 2 min.

4. Add 4 μl of a 100 mM solution of ADP. Allow the chart to run long enough to observe the transition back to the slow rate of respiration when the ADP is phosphorylated.

5. Calculate the respiratory control ratio (RCR) as the ratio of the ADP stimulated rate (state 3) to the unstimulated rate (state 4). Calculate the ADP/O ratio as the exact amount of ADP added (400 nmol) divided by the oxygen consumed during the fast phase of respiration, as shown in *Figure 2*.

6

2.4 Steady state measurement of phosphorylation

A more general method to measure P/O ratios is to measure the rate of P_i and O disappearance (13), similar to early methods using the Warburg manometer. Phosphorylation is measured with a 'phosphate trap' of hexokinase plus glucose which regenerates ADP, allowing large amounts of reaction to occur. However, with an oxygen electrode one is limited to the solubility of oxygen in the medium. By this method phosphorylation is allowed to occur until oxygen equals zero, as shown by the O_2 electrode, and then the reaction is sampled to determine the amount of esterified P_i by ^{32}P methods or by chemical determination of P_i, e.g. by reduction of the molybdate complex, see *Protocol 4.*

Protocol 4. Measurement of rate of phosphate utilization and P/O ratio

1. Place the medium (175 mM sucrose, 10 mM glucose, 10 mM K-Pipes pH 6.5, 1 mM $MgCl_2$, 0.1 mM K-EGTA, 2 mM KP_i pH 6.5, 2.5 U hexokinase, 0.5 mM ADP, and substrate, e.g. 10 mM succinate) in an oxygen electrode with 1 mg/ml mitochondria or other phosphorylating preparation, and allow respiration to proceed until all the oxygen is consumed.

2. Deproteinize a sample of the reaction mixture with 7% TCA, centrifuge to remove protein, and place in a tube marked at 10 ml. Prepare a zero time sample (add 7% TCA before addition of mitochondria).

3. Add 1 ml 4% ascorbic acid and 4 ml 0.125% ammonium heptamolybdate plus 0.1 M H_2SO_4 to the sample. Immediately mix the tube, incubate for 2.5 min, and then mix with 4.5 ml of a solution of 0.4% Na citrate, 0.4% Na arsenite, and 0.4% acetic acid.

4. Adjust the volume to 10 ml total and incubate the tube at 60°C for 5 min. Rapidly cool to room temperature and read the absorbance at 700 nm against a blank without P_i. The absorbance is proportional to the P_i concentration. Use the known P_i concentration in the zero time sample to measure the decrease in P_i as oxidative phosphorylation proceeds to form glucose-6-phosphate.

5. Calculate the P/O ratio from the amount of oxygen taken up, and the amount of P_i esterified to glucose-6-phosphate.

Typical values for mitochondrial P/O or ADP/O ratios are 1.0 for ascorbate–O_2 (site 3), 0.50 for succinate–ferricyanide (site 2), 1.5 for succinate–O_2 (sites 2 and 3), and 2.3 for 3-hydroxybutyrate–O_2 (sites 1 and 2 and 3) (1, 13). Traditional values of 1.0 per coupling site were largely the result of the expectation that there would be larger energy losses in the coupling mechanism (e.g. proton permeability of the mitochondrial inner membrane or 'slip' in the proton pumps (e.g. ref. 14)) which would lead to underestimation of the mechanistic site ratio. However, the fractional ratios above were shown to be the mechanistic ratios when quantitative estimates of energy leaks were made (1).

3. Thermodynamic determination of coupling ratios

If a reaction is reversible, the stoichiometry can be determined by measuring the equilibrium constant. If a reaction is not readily reversible, the equilibrium point can not be found because the rate is kinetically controlled, and zero rate may or may not occur near equilibrium, depending on the details of the mechanism. Cytochrome *c* oxidase is an example of an irreversible enzyme (cytochromes can be oxidized by creating a proton gradient but it does not form oxygen gas), and thus it is not possible to determine its equilibrium constant. Sites 1 and 2 in mitochondria are readily reversible, however, and by appropriate manipulation of the redox potential and phosphate potential (ΔG_{ATP}) the equilibrium of oxidative phosphorylation at these sites can be measured.

3.1 Simple one-point calculation

Often a simple calculation of the amount of redox energy over a segment of an electron transfer chain and the maximum phosphate potential made by the system can be very revealing. The minimum span of the chain should be chosen to minimize error from the fact that a system synthesizing ATP is not actually at equilibrium. As an example, consider the energetics of site 2 in whole mitochondria. This coupling region extends from QH_2 to cytochrome *c*, with the proton transport catalysed by the Q cycle of the bc_1 complex. The extent of reduction of Q and cytochrome $c + c_1$ in state 4 was found to be 63% and 30% respectively (15). To calculate the actual ΔE for this section of the chain the following equation is used (all in millivolts):

$$\Delta E = E_m Q + \frac{59}{2} \log \frac{Q}{QH_2} - E_m(c + c_1) - 59 \log \frac{cyt\ c_{ox}}{cyt\ c_{red}}$$

$$= +41 + (-7) - 225 - 22 = -213\ mV.$$

The standard midpoint potentials are from refs 16 and 17, and the E_m for Q is calculated at the external pH, 7.4. To calculate the energy available for phosphorylation:

$$\Delta G_{redox} = 2 \times (23.062\ kcal/V\ mol) = 0.213\ V = 9.8\ kcal/2e$$
$$9.8\ kcal/2e \times 4.1861\ J/cal = 41\ kJ/2e.$$

The significance of this calculation is seen when it is compared with the ΔG_{ATP} under the same conditions (state 4) (18):

$$\Delta G_{ATP} = 16\ kcal/mol = 67\ kJ/mol.$$

The ratio of the two free energies is then the thermodynamic coupling ratio:

$$\text{P/2e thermodynamic} = \Delta G_{redox}/\Delta G_{ATP} = 41/67 = 0.61.$$

It is clear that site 2 can not have a P/O ratio of 1.0.

Perhaps the most difficult aspect of analysing thermodynamics of a respiratory chain is deciding which pH to use for the $E^{0'}$ of hydrogen carriers. The criterion to use is not simply the location of a carrier which is often not clear, as for ubiquinone, but the side on which the scalar protons are formed in the reaction being analysed. If no scalar protons are formed then either internal or external pH can be used since it will not matter in the end. If scalar protons are formed then it matters. For mitochondria, at site 1 the internal pH should be used because the overall reaction is:

$$NADH + H_{in}^{+} + Q \rightarrow NAD + QH_2,$$

at site 2 the external pH should be used because the reaction is:

$$QH_2 + 2c_{ox} \rightarrow Q + 2c_{red} + 2H_{out}^{+}$$

and for site 3 the external pH should be used because the reaction is:

$$2\ c_{red} + \tfrac{1}{2}\ O_2 + 2H_{out}^{+} \rightarrow 2c_{ox} + H_2O.$$

This last equation is not obvious, but can be understood when the overall process of oxidative phosphorylation is considered. When the proton influx through the ATP synthase driven by the electrogenic transmembrane electron flow in cytochrome oxidase is included, the overall proton disappearance occurs outside.

3.2 Detailed studies of the equilibrium of a coupling site

Here the same experiment described above is carried out, but with electron flow occurring in the reverse direction, driven by ATP hydrolysis. Inverted SMP preparations were used because the redox components, succinate–fumarate and NADH–NAD^{+} can be added to directly interact with the dehydrogenases (19, 20), and because added ATP, ADP, and P$_i$ interact directly with ATP synthase, making the equilibrium point more centrally poised than in mitochondria where an additional proton is used to drive transport of ATP out.

(a) 0.5 mg/ml well-coupled submitochondrial particles (SMP) were equilibrated in 250 mM sucrose, 50 mM K Mops pH 7.1, 5 mM MgSO$_4$, 1 mg/ml fatty acid-free bovine serum albumin, 5 mM KCN, succinate (1–5 mM), and fumarate (5–25 mM) in a dual wavelength spectrophotometer (Amino DW-2 or equivalent) at 550–540 nm at 37 °C. Under these conditions cytochrome $c_1 + c$ were fully reduced.

(b) After the spectrophotometer was balanced, and the signal drift-free,

Peter C. Hinkle

mixtures of ATP, ADP and P_i, and $MgSO_4$ to maintain 5 mM Mg^{2+} were added to create a known ΔG_{ATP} from the equation:

$$\Delta G_{ATP} = \Delta G^{0'} + RT \ln\left(\frac{[ADP][P_i]}{[ATP]}\right)$$

where $\Delta G^{0'}{}_{ATP} = -29.0$ kJ/mol (18). The oxidation of cytochrome $c + c_1$, is observed.

(c) $\Delta G'_{succ-c}$ is calculated from the following equation:

$$\Delta G'_{succ-c} = 2\,F\,[E'_{m\ succ} - E'_{m(c + c_1)}]$$
$$+ RT \ln \frac{[fumarate]}{[succinate]} - 2\,RT \ln \frac{[cyt\ c + c_{1ox}]}{[cyt\ c + c_{1red}]}$$

when $F = 96\,487$ c/mol, $R = 8314$ J/mol/deg, T = absolute temperature, $E'_{m(c + c_1)} = 227$ mV, and $E'_{m\ succ} = 17$ mV.

(d) The relationship between ΔG_{ATP} and ΔG_{succ-c} is analysed, taking into account electrogenic and electroneutral fluxes of protons in the Q cycle, as shown in *Figure 3*. The data shows a thermodynamic ATP/2e = 0.89 ± 0.05, and the interpretation of this number is quite interesting. Because electron flow from succinate to cytochrome c is coupled to the electrogenic flux of two protons inward and the formation of two protons from succinate inside (scalar protons in the reaction) the interpretation of the degree of coupling between ATPase and reverse electron flow depends on the relative fraction of Δp which is ΔpH and $\Delta \Psi$. In other words the redox reaction is:

$$succinate + 2\ cyt{-}c_{in}^{ox} + 2H_{out}^+ = fumarate_{out} + 2\ cyt\ c_{in}^{red} + 4H_{in}^+.$$

If Δp were all ΔpH then the equilibrium would be effected by the $4H_{in}^+$, and if Δp were all $\Delta \Psi$ then the equilibrium would be effected by the two charges which cross the membrane electrogenically (as electrons on cytochrome b). *Figure 3* shows two sets of possible theoretical lines, one for H^+/ATP = 2 for F_0F_1 ATPase, and one for H^+/ATP = 3. From other measurements with SMP in this medium it was found that Δp = 65% $\Delta \Psi$ + 35% ΔpH (24) so the data points are entirely consistent with H^+/ATP = 3, and a predicted mechanistic P/2e stoichiometry of 2/3 from the two electrogenic protons in the Q cycle and three electrogenic protons in the ATPase. This is then also consistent with the results of ADP/O measurements in whole mitochondria described above, where H^+/ATP = 3 + 1 from ADP + P_i transport, and P/2e = 2/4 = 0.5 (site 2).

Most coupling sites are not this complex for interpreting thermodynamic results, but this is a good example to illustrate problems that can arise. An alternative way to have analysed the above data would have been to calculate the $E'_{m\ succ}$ based on the internal pH. That would have eliminated the com-

10

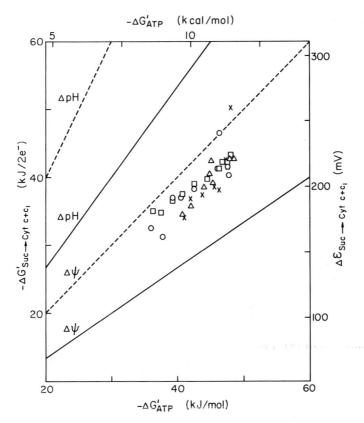

Figure 3. Energetics of reverse electron transfer at site2. The difference in redox potential between succinate and internal cytochromes $c + c_1$ (calculated by using the external pH) is plotted as a function of $-\Delta G'_{ATP}$ of added ATP, ADP, and P_i. The data points indicate [furmarate]/[succinate] ratios of 1 (○), 3 (□), 10 (△), and 25 (×). The lines are the expected equilibrium values of reverse electron transfer if $H^+/ATP = 3$ (solid lines) or $H^+/ATP = 2$ (dashed lines), and if $H^+/2e^- = 2(\Delta\psi)$ or $H^+/2e^- = 4(\Delta pH)$.

plexity in the interpretation and only one predicted line would occur for $H^+/2e$. However, since the succinate is external and the experiment did not involve measurement of internal pH we preferred to use the external pH for $E'_{m\ succ}$ and then introduce ΔpH in the interpretation.

4. Proton transport

The direct measurements of proton transport is more difficult than measurement of ATP synthesis and errors of a factor of two can easily occur. The field is mature enough, however, that a new study should be able to avoid the most significant errors, but as new systems are studied presumably new types of interference will also be found.

11

The main problem is that there is only a small amount of proton flux required to establish Δp across a membrane. If the membrane is impermeant to ions which can serve as counterions for electrogenic proton transport then it is predicted that about 1 nmol H^+/mg protein (an amount that is not possible to measure unambiguously) will charge the electric capacitance of the membrane (1 $\mu F/cm^2$) to 100–200 mV. Therefore it is necessary to provide a counterion for the proton transport which is usually K^+ in the presence of valinomycin but can be tetraphenylphosphonium (+), tetraphenyl borate (−), NO_3^-, or other permeant ions. Under these conditions about 50–100 nmol H^+/mg protein are transported to generate ΔpH, although this amount depends on the system being studied. The use of permeant ions or ionophores such as valinomycin can cause swelling damage to membrane vesicles if the experiment is not designed to avoid extensive transport, and such damage can easily lead to underestimation of correct coupling ratios.

Ratios are usually determined by the pulse method, measuring the transport activity created by a small pulse of substrate such as O_2 or ATP (which is utilized in a few seconds). This is best done with oxygen because the K_m for oxygen reduction is usually very low and the electron transport activity induced over suddenly. Thus a known amount of oxygen can be added and the extent of proton transport measured, allowing a coupling ratio to be calculated. The major error encountered with this method is the masking of some of the proton transport by re-equilibration of permeant acids (P_i, acetate, succinate, or other carboxylic substrates, CO_2/HCO_3) or bases (NH_4/NA_3, Na/H exchange) (21). These permeant molecules rapidly decrease partially the extent of protons measured so that the results look all right but show less transport than they should. The other type of proton leak, electrogenic proton permeability, also decreases the measured proton transport but does so by making the gradient formed decay faster which can be seen during the measurement and then corrected.

Another general principle that has led to the current view of $H^+/2e$ ratios is to focus the measurement on as small a section of electron transfer chain being studied as possible. It is much easier to distinguish between small integers than large integers, and the best situation is to distinguish between zero and some number. This is what happened in the long controversy about mammalian cytochrome oxidase. The assay introduced by Wikstrom (22) used reduced cytochrome c or ferrocyanide as substrate and the observation of any acidification outside mitochondria was strong evidence that cytochrome c oxidase was proton translocating, rather than simply electron translocating. A possible problem with specific site assays is that they often use artificial electron donors or acceptors (phenazine methosulfate, ferricyanide, etc.) which may be much less specific in their interactions with an electron transfer chain than the native substrates.

4.1 Recording pH electrode apparatus

Measurement of protons can be made with a sensitive recording pH electrode apparatus or with a dual wavelength spectrophotometer and pH indicator. The pH indicator has the advantage of a fast response time but may be interfered with by other absorbance changes. The pH meter is very sensitive and can be used in all situations. I originally used an apparatus similar to that described by Mitchell *et al.* (4), with a separate pH electrode and reference electrode. This type of apparatus is designed to minimize the noise generated by salt gradients in the reference electrode connection to the reaction medium. However, it is hard to clean and is sensitive to electrostatic pick-up, so I later changed to the simpler set-up shown in *Figure 4*. The combination electrode is much less sensitive to pick-up of electrostatic noise, but requires high salt concentration in the reaction medium to avoid noise. The electrode should be adjusted to be as close to the stirring bar as possible to improve the response time. The response times of pH electrodes are dependent on frequent cleaning of the surface with a tissue and detergent solution, the stirring efficiency near the electrode surface, and the buffer concentration. Since the buffer concentration can not be high because it decreases extent of the pH change caused by proton transport, a compromise must be made between sensitivity and response time which depends on the details of the experiment.

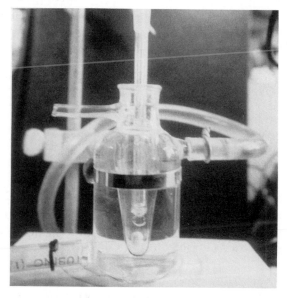

Figure 4. pH electrode cell made by local glass blower. A similar cell can be made from a Plexiglass box and a test-tube. A combination pH electode is held just above the stirrer, and N_2 or argon gas is gently blown over the solution.

The electronics used can be important. Most commercial pH meters are not ideal for recording because they have a slow time constant at the input and may not allow convenient interface with a recorder to provide 0.1 pH full-scale. The requirements for a recording pH meter are low noise, high response time, and high sensitivity, while absolute accuracy is not important since calibrations of the amount of protons are made during the experiment. A simple electrometer circuit using a field effect transistor operational amplifier powered by batteries or a simple ±15 volt supply can easily be assembled from locally available parts.

4.2 H$^+$/O ratio

Because extensive respiration in the presence of valinomycin and potassium ions causes swelling of mitochondria the valinomycin should be added after the system is anaerobic, which can usually be observed by a small change in the pH electrode trace. To avoid excessive respiration the reaction medium should be largely oxygen-free at the start. This is accomplished by placing about 30 ml of medium in a 50 ml syringe and clamping the syringe facing upward with about 10 ml of air at the top of the chamber. N_2 is then bubbled through the medium with a fine plastic tube through the needle connector for about 10 min. The N_2 phase is then expelled and a needle attached. This syringe of anaerobic medium can be used to fill the cell throughout the day. Some drift in pH is caused by the diffusion of CO_2 out of the medium causing a drift toward basic pH, even without mitochondria. If this occurs the medium should be briefly boiled before being bubbled in the syringe. An oxygen pulse experiment is described in *Protocol 5*.

Protocol 5. Oxygen pulse experiment to determine H$^+$/O ratios

1. Add 2 ml of anaerobic medium to the cell (see text). Turn on the N_2 stream above the medium and add 5–10 mg of mitochondria, followed by substrate (2 mM succinate), appropriate inhibitors (0.1 μM rotenone in ethanol and 0.2 mM *N*-ethylmaleimide), and any additions of 10 mM HCl or KOH to bring the medium to the exactly desired pH. Wait until the pH recording trace is relatively drift-free.

2. When a stable recording and anaerobiosis is reached (in 2–5 min) add valinomycin (1 μM from ethanol solution) and allow the resulting acid extrusion, caused by oxygen in the valinomycin solution, to decay (2–5 min).

3. Rapidly add an oxygen pulse of 5–20 μl air-equilibrated medium without buffer and record the resulting acid extrusion at about 10 cm/min chart speed. Decrease the chart speed after a few minutes to allow the final end-point and possible drift during the experiment to be estimated.[a]

---

4. Measure the sensitivity of the system to protons by the addition of a small aliquot of anaerobic[b] 10 mM HCl to change the pH by about 0.02 units.

5. Calculate the H^+/O ratio from the moles of H^+ formed by the oxygen pulse and the moles of O added. Calculate the oxygen added from the known concentration of air-equilibrated solutions, as described in section 2.2. Estimate the extent of H^+ transport by plotting the decay of the gradient formed on semi-logarithmic graph paper as the extent above the end-point.[c]

[a] After complete decay of the gradient another O_2 pulse can be given.
[b] HCl can be made anaerobic by bubbling with N_2 in a test-tube covered by Parafilm, and the additions drawn with a microsyringe through the Parafilm.
[c] Various attempts have been made to make more accurate extrapolations or corrections to obtain the true amount of proton transport (e.g. ref. 22), but they are not recommended. The best way to correct for interference is to discover the cause and eliminate it as much as possible.

4.3 H^+/ATP ratio

The H^+/ATP ratio can be determined by the same pulse methods as the H^+/O ratio, except that the results have never been satisfactory because the K_m for ATP is quite high and instead of a square wave of activity one obtains a sawtooth wave which decays slowly to zero. In addition the ATPase reaction causes a net pH change, although this can be avoided by working at about pH 6.1 at 2 mM Mg^{2+} where the reaction is pH neutral (22). In any case the extent of ATP hydrolysis must be measured because adenylate kinase is always present to some extent and the added ATP is hydrolysed to AMP with complex kinetics. It is probably best to measure P/O ratios and H^+/O ratios and then calculate the H^+/ATP ratio from the others. In this way the H^+/ATP ratio for F_1F_0-type ATP synthases is probably 3, and the ratio for synthesis and transport of ATP out of mitochondria is probably 4.

Measurement of the H^+/ATP ratio by the thermodynamic equilibrium method has been attempted many times and gives 3 for F_1F_0 (e.g. ref. 24), but also 3 for synthesis and transport in mitochondria (25). This is probably due to an incomplete equilibrium in mitochondria where the extreme ADP concentration is only a few micromolar, or to systematic errors in Δp measurement, which is difficult to prove error-free.

References

1. Hinkle, P. C., Kumar, M. A., Resetar, A., and Harris, D. L. (1991). *Biochemistry*, **30**, 3576.
2. Slater, E. C. (1967). In *Methods in enzymology* (ed. R. W. Estabrook and M. E. Pullman), Vol. 10, pp. 19–29.
3. Hinkle, P. C. and Yu, M. L. (1979). *J. Biol. Chem.*, **254**, 2450.

4. Mitchell, P., Moyle, J., and Mitchell, R. (1979). In *Methods in enzymology* (ed. S. Fleischer and L. Packer), Vol. 55, pp. 627–40. Academic Press, New York.
5. Reynafarje, B., Brand, M. D., Alexandre, A., and Lehninger, A. L. (1979). In *Methods in enzymology* (ed. S. Fleischer and L. Packer), Vol. 55, pp. 640–56. Academic Press, New York.
6. Estabrook, R. W. (1967). In *Methods in enzymology* (ed. R. W. Estabrook and M. E. Pullman), Vol. 10, pp. 41–7. Academic Press, New York.
7. Chappell, J. B. (1964). *Biochem. J.*, **90**, 225.
8. Lemasters, J. J. (1984). *J. Biol. Chem.*, **259**, 13123.
9. Thayer, W. S., Tu, Y. L., and Hinkle, P. C. (1977). *J. Biol. Chem.*, **252**, 8455.
10. Chance, B. and Williams, G. R. (1955). *J. Biol. Chem.*, **217**, 383.
11. Lamprecht, W. and Trantschold, I. (1974). In *Methods of enzymatic analysis* (ed. U. H. Bergmeyer), pp. 2101–10. Academic Press, New York.
12. Krishnamoorthy, G. and Hinkle, P. C. (1984). *Biochemistry*, **23**, 1640.
13. Stoner, C. D. (1987). *J. Biol. Chem.*, **262**, 10445.
14. Pietrobon, D., Luvisetto, S., and Azzone, G. F. (1987). *Biochemistry*, **26**, 7339.
15. Muraoka, S. and Slater, E. C. (1969). *Biochim. Biophys. Acta*, **180**, 227.
16. Urban, P. F. and Klingenberg, M. (1969). *Eur. J. Biochem.*, **9**, 519.
17. Ohnishi, T., Wilson, D. F., Asakura, T., and Chance, B. (1972). *Biochem. Biophys. Res. Commun.*, **46**, 1631.
18. Rosing, J. and Slater, E. C. (1972). *Biochim. Biophys. Acta*, **267**, 275.
19. Rottenberg, H. and Gutman, M. (1977). *Biochemistry*, **16**, 3220.
20. Scholes, T. A. and Hinkle, P. C. (1984). *Biochemistry*, **23**, 3341.
21. Brand, M. D., Reynafarje, B., and Lehninger, A. L. (1976). *J. Biol. Chem.*, **251**, 5670.
22. Wikström, M. (1977). *Nature*, **266**, 271.
23. Thayer, W. S. and Hinkle, P. C. (1973). *J. Biol. Chem.*, **248**, 5395.
24. Berry, E. A. and Hinkle, P. C. (1983). *J. Biol. Chem.*, **258**, 1474.
25. Woelders, H., van der Zaude, W. J., Colen, A. M. A. F., Wanders, R. J. A., and van Dam, K. (1985). *FEBS Lett.*, **179**, 278.

2

Membrane permeability and transport

RALF T. VOEGELE, EMMALEE V. MARSHALL, and
JANET M. WOOD

1. Introduction

The limited permeability of biological membranes to large and/or hydrophilic solutes allows cells to modulate their own composition and to base their energy metabolism on transmembrane electrochemical potential gradients. Our current understanding of those phenomena, which were first placed within a comprehensive theoretical framework by Peter Mitchell (1), is based on measurements of membrane structure, of membrane permeability, and of transmembrane solute fluxes mediated by enzymes denoted as transporters. Photosynthetic and respiratory electron transport chains and ion-motive adenosine triphosphatases (ATPases) mediate creation of the membrane potential ($\Delta\psi$), formation of transmembrane electrochemical potential gradients ($\Delta\mu_1$), and coupling of those gradients to the synthesis and hydrolysis of ATP. The membrane potential, solute gradients, ATP, and phosphoenol-pyruvate (PEP) serve as the energy supplies powering many other transport reactions. Methodologies for the estimation of $\Delta\psi$ and of the protonmotive force ($\Delta\mu_{H+}$) are described in Chapter 3 of this volume. This chapter describes methodologies for the measurement of passive and transporter-mediated transmembrane solute fluxes. It does not describe methods specifically designed for the study of membrane channels. Detailed descriptions of theories, procedures for the preparation of cells, organelles, and membrane vesicles, and of methods for transport measurements can be found in other books in the Practical approach series and in the Methods in enzymology series of books.

1.1 Experimental systems

Transmembrane solute fluxes can be measured within a variety of experimental systems. They include (planar) black lipid membranes and topologically closed, model membrane vesicle systems such as liposomes (phospholipid vesicles), proteoliposomes (vesicles reconstituted from phospholipid and protein), and biological membrane vesicles (both right-side out and everted with respect to their orientation in the intact cell), as well as intact organelles, cells, and tissues. All of these systems provide at least two compartments separated by at least one phospholipid-based membrane. This chapter describes methodologies designed for the measurement of solute fluxes across the membranes of biological membrane vesicles, organelles, and cells. The flow dialysis technique (section 2.1.3), electrochemical detection of ion fluxes (section 2.2), and methods based on integrated or dynamic light scattering (section 2.3) are also applicable to (proteo)liposome systems, and the procedure for controlled aggregation of (proteo)liposomes described recently by Tortorella *et al.* (2) has the potential to extend application of filtration- and sedimentation-based methodologies (sections 2.1.1 and 2.1.2) to those systems. Although the specific examples cited below are drawn from our experience in studying transmembrane solute fluxes in Gram-negative bacteria (and in membrane vesicles derived from them), these methods are applicable with minimal modification to other experimental systems.

1.2 The transport vocabulary

Transmembrane solute fluxes can be differentiated in several respects, including whether they are mediated by a catalyst, whether they are linked to an exogenous energy supply (an energy source other than a pre-existing electrochemical potential gradient of the solute under consideration), whether they create (or enhance) the transmembrane chemical potential (or concentration) gradient of that solute, and whether they are associated with its covalent modification. **Passive flux** is the unmediated movement of a solute across a phospholipid-based membrane down its own electrochemical potential gradient. Following the general practice of enzymology, solutes whose flux is mediated by a transporter (a protein catalyst) are denoted as **substrates**. In contrast to transporters which, like enzymes, possess specific substrate binding sites, **channels** provide transmembrane aqueous pathways through which solutes diffuse down their electrochemical potential gradients. Since they do not include specific substrate binding sites, most channels show relatively little solute specificity.

Transport is the mediated flux of a substrate across a phospholipid-based membrane, without covalent modification. It includes both **facilitated diffusion** (in which a transporter catalyses movement of a substrate across a membrane down its own electrochemical potential gradient) and **active transport** (in which a transporter couples movement of a substrate across a membrane up its own electrochemical potential gradient with another, energy-yielding reaction). Active transport may be coupled to the energy-yielding bond cleavage reactions (**primary active transport**) or to the dissipation of the transmembrane electrochemical potential gradient of a co-substrate (**secondary active transport**). Secondary active transport (or co-transport) is described as **symport**

or **antiport** depending upon whether the co-transported substrates move across the membrane in the same or opposite directions, respectively. **Group transloca-tion** is transmembrane flux which is obligatorily linked to chemical modification of the substrate. It is distinguished from other flux mechanisms by the fact that it does not, in the same sense, create a transmembrane solute gradient.

Uptake and **efflux** are empirical terms which describe solute movements without specifying the flux mechanism or the fate of the translocated molecules. Whereas uptake or efflux measurements which incorporate solute metabolism may be useful in some situations, rigorous analysis of flux mechanisms often requires that solute fluxes be resolved from other reactions. To determine whether measurements of transmembrane flux incorporate solute modifica-tion, the product(s) of the transport reaction must be recovered and analysed. This is most easily accomplished through the use of radiolabelled substrates (see section 2.1).

1.3 Applications of theoretical models for membrane permeability and transport

In attempting to compare the permeabilities of diverse membrane systems and to elucidate solute flux mechanisms, researchers have developed a variety of theoretical models. These models are used during analysis of data describ-ing both the kinetics and the thermodynamics of transport. In their most rudimentary form, such analyses yield descriptive parameters that are useful in distinguishing one flux pathway from another or in detecting changes in transporter function or environment. On a more sophisticated level, kinetic and thermodynamic parameters may reveal detailed flux mechanisms. A specific protein can only be linked with an observed transport activity by examining the pure protein reconstituted in proteoliposomes, however.

1.3.1 Kinetic models

In spite of their evident complexity, transmembrane solute fluxes can be described empirically in terms of system-specific permeability coefficients (K_P) that are analogous to the diffusion coefficients (K_D) that describe solute fluxes as a function of concentration gradients in solution:

$$v_0 = K_P \, \Delta s$$

where v_0 is the initial rate of flux of a solute down the transmembrane concentration gradient Δs. This simple formulation does not take into account effects of membrane potential on flux rates for charged solutes, however. Radiotracer experiments permit unidirectional fluxes to be moni-tored even in the presence of counterfluxes of the same (unlabelled) solute. For passive flux, K_P is determined by factors such as solute and membrane structure, temperature, and membrane area.

Mediated transmembrane solute flux is characterized by deviations from the linear dependence of solute flux (v_0) on solute concentration gradient (s) illustrated above. For mediated flux, the dependence of (initial) rate (v_0) on the concentrations of transporter (or enzyme) (e_0) and substrate (s) can be described in terms of the Michaelis–Menten parameters K_M, k_{cat} and the derivative parameter V_{max} ($k_{cat} e_0$) as for other enzyme catalysed reactions:

$$v_0 = \frac{V_{max}\, s}{(K_M + s)}.$$

In many membrane systems, transporters with overlapping substrate specificities function in parallel with passive substrate flux and/or binding reactions. While passive fluxes and binding reactions are often negligible in comparison to active transport, they can become significant contributions to measured fluxes under some circumstances. Data describing the kinetics of multiple, simultaneous transport and competing processes can be analysed in terms of equations with the form:

$$v_0 = \frac{V_{max1}\, s}{(K_{M1} + s)} + \frac{V_{max2}\, s}{(K_{M2} + s)} + Ks$$

in which the first and second terms on the right side of the equation describe substrate transport mediated by two populations of transporters with different kinetic properties, and the third term on the right side of the equation describes passive flux, very low affinity-mediated flux, and/or substrate binding.

Although linearizing transformations may be useful in identifying deviations from simple saturation kinetics for transport, as for other enzyme catalysed reactions, non-linear regression analysis is the preferred method for derivation of kinetic parameters. It avoids statistical distortions, readily permits evaluation of the fit of experimental data to diverse models, and yields estimates of error in the deduced parameters. Many commercially available computation programs now support non-linear regression. Initial rate kinetic analysis coupled with stepwise genetic deletion of transporters can greatly facilitate the resolution of multiple transport components (3). Detailed examinations of substrate specificity can also be useful in this regard (4). Testing for the occurrence of multiple, parallel flux pathways is important since they are common and difficult to distinguish, on the basis of kinetic data alone, from the activity of a single transporter population which does not show simple Michaelis–Menten behaviour.

Application of the Michaelis–Menten equation is based on the assumption that reaction product concentration(s) remain(s) negligible during the initial rate measurement period. Violations of that assumption are particularly likely during transport reactions which result in concentrative uptake of an unmodified substrate. Such violations can be avoided and the apparent initial rate period of the reaction prolonged by ensuring that transport is followed by intracellular or intravesicular modification of the substrate, as long as the reaction products are not themselves lost through efflux during the initial rate measurement or active as transport inhibitors. Rapid uptake measurements, accomplished within milliseconds after initiation of the uptake reaction, may also be used to obtain initial uptake rates (see *Protocol 3*).

Analyses of initial rate data according to the procedure described above are also based on the assumption that co-substrate concentrations and the membrane potential (where pertinent) are held constant. In the case of active transport processes, co-substrates may include co-transported solutes or compounds such as ATP or PEP whose hydrolysis powers substrate flux. Control of co-substrate concentrations and the membrane potential in whole cell systems may be difficult, though those quantities can be independently estimated. Membrane vesicle systems usually afford greater opportunity to systematically vary transport reaction conditions. The dependence of transport rates on co-substrate concentration and on the membrane potential has been modelled for both primary (5) and secondary (6) active transport processes. Systematic variation of the concentrations of potential coupling ions or of other co-substrates and of the membrane potential can be employed, for example, to confirm their identities and to determine substrate binding sequences and interdependencies (7, 8).

Efforts to derive mechanistic details from transport kinetic data were long delayed by the absence of experimental systems in which the activities of homogeneous transporter populations could be measured without interference from competing reactions. The application of genetic and molecular biological techniques has now led to deletion or amplification of genes encoding specific transporters and to amplification of transporter gene expression. This has facilitated both purification and reconstitution of transporters in proteoliposomes and derivation of biological membrane systems in which a single transporter population is dominant. In addition, it has led to the creation and characterization of transporter variants. Application of kinetic methodologies to those new experimental systems is now yielding long-awaited mechanistic information (9).

1.3.2 Thermodynamic models

In practical terms, measurements of transport thermodynamics may be used to distinguish active transport from passive transmembrane flux, facilitated diffusion, or group translocation, to identify the driving force for active transport, and to further define transport mechanisms. The electrochemical potential gradient of a solute (S) across a membrane is defined as:

$$\Delta\tilde{\mu}_s = nF\Delta\psi + 2.3RT \log \frac{[S]_i}{[S]_0}$$

(units of joules)

$$\frac{\Delta\tilde{\mu}_s}{F} = n\Delta\psi - \frac{2.3RT}{F}\Delta pS$$

(units of volts)

21

where n is the net charge on the solute, F is the Faraday constant (96.5 kJ/V), $\Delta\psi$ is the membrane potential (in volts), R is the gas constant (8.314 J/mol K), T is the temperature (K), $[S]_i$ and $[S]_0$ are the concentrations of the solute inside and outside the cell or vesicle, respectively, and ΔpS is equal to pS_0 minus pS_i, the transmembrane solute gradient. During passive flux and facilitated diffusion solutes distribute across the membrane according to the above equation (see *Table 1*). Those processes can, therefore, result in the creation of a solute concentration (chemical potential) gradient if a charged solute redistributes in response to a pre-existing membrane potential ($\Delta\psi$). Many weakly acidic solutes are membrane impermeant only when ionized. Such solutes spontaneously redistribute in response to pre-existing proton gradients (ΔpH). Thus, to unambiguously determine whether active transport is occurring, the transmembrane distribution of free solute must be determined and the above equation must be used to determine whether $\Delta\tilde{\mu}_s$ differs from zero. Solute distribution must be measured under equilibrium (or at least steady state) conditions and the solute must be neither bound nor covalently modified following transmembrane flux. If the solute is charged, the membrane potential must be independently determined and included in the calculation. With the exception of the passive flux or facilitated diffusion of uncharged solutes, most transmembrane fluxes yield transmembrane chemical potential gradients.

By correlating measured substrate electrochemical potential gradients with those of putative co-substrates, it has been possible to obtain evidence for both co-substrate identity and coupling stoichiometry (10). For example if, over a variety of conditions, the magnitude of the substrate electrochemical

Table 1. Definition of terms describing solute fluxes

Term	Flux catalysed?	Solute modified?	Solute concentration gradient created?
Passive flux			
Uncharged solute	No	No	No
Ionic solute	No	No	Yes, if $\Delta\psi$
Weakly acidic solute	No	No	Yes, if ΔpH
Transport			
Active transport	Yes	No	Yes
Facilitated diffusion			
Uncharged substrate	Yes	No	No
Ionic substrate	Yes	No	Yes, if $\Delta\psi$
Weakly acidic substrate	Yes	No	Yes, if ΔpH
Co-transport, symport, and antiport	Yes	No	Yes

22

potential gradient is consistently twice the magnitude of the proton electro-chemical potential gradient, it is plausible that the transport phenomenon under study is proton/substrate co-transport with a coupling stoichiometry of two protons per substrate molecule. Such conclusions should be drawn with care, however, since the transmembrane electrochemical potential gradients of many species are linked to one another through co-transport processes (e.g. ref. 11). Similarly, for primary active transport, the electrochemical potential gradient of a substrate can be correlated with the chemical potential of the putative coupling reaction (for example, the phosphorylation potential in the case of ATP-driven transport).

2. The measurement of permeability and transport

As indicated by the above discussion, measurements of transmembrane solute flux (kinetic measurements) and distribution (thermodynamic measurements) may be used to distinguish passive from mediated transmembrane solute flux, and active transport from facilitated diffusion. For mediated fluxes, such measurements can define transporter multiplicity and concentration, kinetic parameters, substrate specificity, co-substrate participation, and stoichiometry as well as other, more sophisticated aspects of transport mechanisms. The best methodology for the detection and estimation of a particular solute flux or distribution depends, of course, upon the specific question asked and the particular properties of the experimental system under study.

2.1 Radioisotopic methods

Although some transport measurements have been based on post-flux solute modification, such methods are clearly not applicable when transmembrane flux, alone, is to be detected (for example during estimation of solute distribu-tion). For practical reasons, most measurements of solute flux or distribution are based upon solute uptake. Application of uptake methodology to vesicles that are everted in orientation with respect to the cells from which they were derived can also be used to simulate solute efflux. For each method, cells or vesicles are incubated for a defined period with a solution containing the radiolabelled solute. Various techniques are then used to separate the cells or vesicles from the solute-containing medium (filtration (*Protocols 1, 2,* and *3*), sedimentation (*Protocol 4*), or exclusion chromatography) or to estimate the concentration of solute remaining in that medium (flow dialysis) (*Protocol 5*). The radioactivity taken up (or not taken up) is then estimated with an appropriate radioisotope detection method.

While simplest, best adapted for kinetic measurements and most commonly used, filtration is also the most perturbing of the available separation tech-niques. Some vesicles can not be recovered by filtration or by sedimentation. The advantages and disadvantages of the filtration and sedimentation method-ologies have been discussed by Bakker (12). Exclusion chromatography, in some cases accelerated by centrifugation, can be used to address that problem. Flow dialysis can be applied when cell or vesicle systems are labile or refractory to filtration or sedimentation or when investigators have reason to think that those techniques will perturb transmembrane solute distribution. Flow dialysis is better adapted for thermodynamic than for kinetic measurements, however. Each of these methodologies is, in principle, applicable for the measurement of passive flux, facilitated diffusion, and active transport. In practice, equilibrative fluxes (both passive and facilitated) may be difficult to detect as they are often both rapid and small in magnitude. That problem can sometimes be addressed by coupling transmembrane flux to a secondary, 'trapping' reaction. An alternative approach may be to exploit the osmotic consequences of such fluxes by monitoring vesicle swelling or shrinkage (see section 2.3).

2.1.1 Filtration methodologies

Most commonly, suspending media are removed by trapping cells or vesicles on filters and washing them with the help of a vacuum manifold. Trapping, which must be highly efficient and reproducible, may be accomplished through sieving and/or through binding to the chosen filters. These methodologies permit estimation of uptake after incubation times as short as five seconds (standard filtration methodologies, *Protocols 1* and *2*) or even milliseconds (experiments performed with a rapid filtration apparatus) (*Protocol 3*).

Protocol 1. Solute uptake by whole cells using the standard filtration methodology

Equipment and reagents

- Manifold/filter apparatus (Hoefer Scientific Instruments) with vacuum pump (a side arm flask with a filter support is sufficient, but manifolds that accept multiple filters are convenient and commercially available)
- Temperature controlled water-bath
- Filters (for *Escherichia coli*): cellulose nitrate/cellulose acetate membrane filters, pore size 0.45 μm, diameter 25 mm, plain surface (catalogue no. HAWP 025 00, Millipore Corporation)
- Cell suspension in transport assay buffer (see step 1).
- Solution of cellular energy source (e.g. D-glucose to yield a final concentration of 10 mM)
- Solution of radioisotopically labelled solute with known specific radioactivity (see *Hint* (e))
- Wash buffer, at water-bath temperature (see *Hint* (b))
- Micropipettors and/or Hamilton syringes
- Radioisotope monitor, scintillation vials, scintillation fluid, scintillation counter

Method

1. Grow cells in desired medium to the desired growth stage. Wash them at least twice with medium devoid of the solute whose uptake is to be measured. Adjust the cell suspension to the desired optical density (for *E. coli*, 0.1 to 1.5 at wavelength 578 nm) or protein concentration (for *E. coli*, 0.015 mg/ml to 0.2 mg/ml).

2. Mix the cell suspension in a reaction tube with the uptake energy source (if required) and incubate for a specific time in a temperature controlled water-bath.[a] Aerate by shaking or bubbling with air.

3. Place filters (pre-wet with washing buffer) onto manifold and turn on vacuum pump.

4. Start uptake reaction by adding labelled solute. At desired time points (e.g. 15 sec, 30 sec, and 45 sec), remove aliquots (typically 150 μl) of the cell suspension, place them on filters, allow filtration to occur, and wash with approximately 10 ml of wash buffer.

5. When samples have been applied to all filters on the manifold, remove the filters carefully, place them in scintillation vials, add scintillation fluid, and count.

Helpful hints

(a) Pre-wetting of cellulose nitrate/cellulose acetate filters is important. It facilitates removal of the separation sheets and improves the manifold vacuum.

(b) Some experimenters quench the uptake reaction by diluting the assay mixture with wash solution or a transporter inhibitor (e.g. $HgCl_2$) prior to filtration.

(c) Save some of the cell suspension to determine its protein concentration.

(d) Always include negative controls such as assay mixtures from which the cell suspension or the energy supply has been omitted or for which the cell suspension has been boiled prior to the assay. Negative control values are important as zero time uptake estimates in defining the uptake time course, and as zero accumulation estimates in defining steady state solute distributions.

(e) Apply an aliquot of the radiolabelled substrate solution to a filter (without evacuation), transfer the filter to a scintillation vial, and count the retained radioisotope in order to determine the correspondence between counts per minute (c.p.m.) detected by the scintillation counter and solute molar quantity (the 'operational' specific radioactivity (*SR*) in moles per c.p.m.). This estimate will, of course, require knowledge of the precise chemical concentration of the solute in the solution.

(f) Adjust the solute specific radioactivity, solute concentration, and cell volume added to each assay mixture to ensure that more than 1000 c.p.m. are recovered per filter. For initial rate measurements, less than 10% of the available radioisotope should be taken up so that the extracellular solute concentration remains essentially constant.

Ralf T. Voegele et al.

Protocol 1. *Continued*

(g) This methodology can be adapted to determine whether the solute is modified after uptake. The wash buffer is replaced with a solution that extracts cellular metabolites (ice-cold water, for example). A test-tube is placed within the manifold to capture the wash solution passing through each filter. The collected extracts are lyophilized and then analysed appropriately (by thin-layer or high-pressure liquid chromatography, for example).

[a] A typical assay mix would include cell suspension (475 μl), energy source (5 μl), and substrate (20 μl) for a total volume of 500 μl.

A very similar methodology can be applied to cytoplasmic membrane vesicles if they can be trapped by filtration and if solutes do not redistribute across the membrane during filtration.

Protocol 2. Solute uptake by membrane vesicles using the standard filtration methodology (13)

Equipment and reagents

- See *Protocol 1*
- Filters[a]
- Assay buffer: 50 mM potassium phosphate, 10 mM MgSO$_4$, pH 6.6
- Wash buffer: 0.1 M LiCl
- Solution of respiratory energy source (e.g. Li D-lactate) to yield a final concentration of 20 mM

Method

1. Use freshly prepared vesicles or vesicles that have been rapidly thawed after storage at −70°C. Dilute vesicles to the appropriate protein concentration (3.0–3.5 mg/ml) in assay buffer and keep them on ice until use.

2. Prepare assay tubes containing assay buffer and keep them on ice.

3. Add vesicles to the assay mixture and bring it to the assay temperature in a temperature controlled water-bath.[b]

4. Add energy source to the assay mixture, vortex, and incubate further.

5. Complete the assay as described above for whole cells, using longer time intervals if necessary.

[a] If cellulose nitrate/cellulose acetate filters of 0.45 μm pore size are employed, it may be desirable to pre-wet them with 0.1 M LiCl to ensure binding of the vesicles to the filter. With glass fibre filters no pre-wetting is required. If cellulose nitrate/cellulose acetate filters with smaller pore diameter are used (e.g. 0.1 or 0.2 μm), it may be necessary to reduce the vesicle concentration to achieve rapid filtration.
[b] A typical assay mix would include assay buffer (330 μl), membrane vesicle suspension (50 μl), energy source (100 μl) and substrate (20 μl) for a total volume of 500 μl.

A special apparatus has been designed to permit the estimation of initial uptake rates over very short (millisecond) incubation periods (14). A cell or vesicle sample is applied to a filter and the radiolabelled solute is passed through the filter for a defined time. The contact between them is then terminated and only the solute that has been taken up by the cells or vesicles during that time remains on the filter. Different time points result from separate experiments performed using separate cell or vesicle suspension aliquots and filters. This is in contrast to the standard filtration method, in which several time points can be estimated using one reaction mixture.

Protocol 3. Solute uptake using the rapid filtration
methodology

Equipment and reagents

- Filtration apparatus (Bio-Logic RFS-4, Intracel Ltd.) (14)
- See *Protocol 1*

Method

1. Prepare a cell or vesicle suspension as described in *Protocol 1* or *2* and a large volume (100 ml) of radiolabelled solute solution.

2. Dilute cells or vesicles to a suitable concentration that can be evenly layered onto the filter and still allow an adequate flow rate.

3. Put a filter onto the filter holder, slide the filter holder under the syringe body, and then carefully pour the radioactive solute solution into the syringe. Set the controller to zero. Slowly, lower the plunger until it is just at the top of the syringe barrel. Slide the filter holder back out and remove the filter.

4. Set the desired flow rate and time on the controller. Place a pre-wet filter onto the filter holder. Place the adaptor on top and carefully layer the sample evenly on the filter.

5. Turn the vacuum on. Slide the filter holder under the body of the syringe and press the start button on the controller.[a]

6. Slide the filter holder out, transfer the filter to a scintillation vial, and add scintillation fluid. Count.

---

[a] This will initiate and terminate the flow of substrate solution through the filter. The controller will display the actual time and volume that the apparatus used. If the actual time shown is not the time set, the flow rate may need to be adjusted to allow for adequate flow and hence to obtain the desired time.

When solute uptake is measured using radiolabelled compounds, the primary data include the quantity of radioactivity (counts per minute or c.p.m.)

trapped on each filter and the concentration of the cell or vesicle suspension used for the uptake assay (*C*, usually in terms of dry weight or quantity of protein per unit volume). In addition, the following experimental parameters must be defined:

V_T the total volume of the uptake assay mixture
V_A the volume of the aliquot of the total uptake assay mixture applied to each filter
V_C the volume of the cell or vesicle suspension added to the uptake assay mixture
SR the 'operational' specific radioactivity (or correspondence between moles of solute and c.p.m. detected by the scintillation counter) of the radiolabelled solute (defined in *Protocol 1, Hint* (d)).

The quantity of solute taken up by the cells (*M*) can then be computed as:

$$M = \frac{(\text{c.p.m.}) \, V_T}{(SR) \, V_A \, V_C \, C}$$

The units of *M* are moles of solute per milligram of cells/vesicles (as dry weight or protein). This value may require correction for background quantities (see *Protocol 1, Hint* (c)). In the case of kinetic measurements, *M* should be divided by the interval after initiation of the reaction at which filtration occurred to yield a rate. The uptake time course should be defined to determine the time interval appropriate for initial rate measurements. Uptake should then be determined at a minimum of three time points to define a rate. If the specific radioactivity of the solute is held constant within an experiment, background quantities that are time-independent during the initial rate period (usually, for example, binding of solutes to cells and filters) need not be independently subtracted when rates are determined in this way, but attention should always be paid to background values as they may provide insight regarding the assay system.

2.1.2 Sedimentation methodology

Sedimentation can be used to estimate slow uptake rates and the accumulation of solutes whose transmembrane distribution would be perturbed during filtration (15, 16). In this case cells are separated from their surrounding medium by centrifugation through a cushion of silicon oil. This method is not normally applied to vesicles since the readily available sedimentation systems are not capable of effecting their sedimentation. Aggregation of vesicles using the procedure of Tortorella *et al.* (2) may address that problem, however. *Protocol 4* describes determination of the transmembrane distribution of a ^{14}C-labelled solute in comparison with that of tritiated water. For kinetic determinations, the tritiated water is omitted and special care is taken to ensure that each cell pellet is fully recovered.

Protocol 4. Solute uptake by the sedimentation methodology

- Temperature controlled water-bath
- Microcentrifuges (at least two, with rapid acceleration and minimum sedimentation capability of 12 000 g, preferably with 90° fixed angle rotors, e.g. Microfuge E, Beckman Instruments Inc.)
- Microcentrifuge tubes
- Vacuum pump with solvent trap
- Cell suspension (prepared as described in *Protocol 1* except that a higher cell density is required) (for *E. coli*, an optical density (578 nm) of 1.0 to 3.0)
- Silicon oil (for bacteria, Wacker Chemie, silicon oil AR200, $\delta = 1.04$ at 25°C, or equivalent oil)
- Pipettes, automatic pipettors, drawn-out Pasteur pipettes
- Tritiated water
- Solution of ^{14}C-labelled solute with known specific radioactivity
- Solution of energy source
- 0.4 M NaOH
- Scintillation vials, scintillation fluid, scintillation counter

Method

1. Prepare cell suspension as described in *Protocol 1*. Mix an aliquot of the cell suspension with tritiated water to give a final radioactivity of 0.5 mCi/ml. Add a transport energy source (if necessary).[a]

2. Start the reaction by adding the ^{14}C-labelled solute.[b]

3. At the desired time points (e.g. 1 min, 3 min, and 5 min) transfer aliquots of the cell suspension (usually 0.5 ml to 1.0 ml) onto a 200 μl cushion of silicon oil in a 1.5 ml microcentrifuge tube. Place the tube in a microcentrifuge and sediment for 2 min at 12 000 g.[c]

4. After centrifugation, transfer a sample of the upper phase (usually 50 μl) to a 5 ml scintillation vial containing 0.75 ml of 0.4 M NaOH (this sample gives the '(^{14}C-substrate/^{3}H$_2$O) outside' ratio). Carefully remove the rest of the upper phase and most of the silicon oil (lower phase) with a drawn-out Pasteur pipette connected to a vacuum pump and trap. Cut off the bottom of the Eppendorf tube containing the pellet and place it in a 5 ml scintillation vial containing 0.75 ml of 0.4 M NaOH (this sample gives the '(^{14}C-substrate/^{3}H$_2$O) inside' ratio).

5. Mix all samples (especially the samples containing the pellet) and incubate them at 37°C for 1 h. Add 5 ml of scintillation cocktail to each sample, mix by repeated inversion, hold in the dark for at least 1 h, and count in a scintillation counter set appropriately for dual-label counting.

Helpful hint

This methodology can be adapted to determine whether the solute is modified after uptake. The tritiated water is omitted from the sample and a small volume (50–100 μl) of a 10% (w/v) trichloroacetic acid (TCA) solution is placed in the microcentrifuge tube before the silicon oil is

Protocol 4. *Continued*

carefully layered on top of it (the percentage of the TCA solution may have to be adapted if the density of the silicon oil is altered). After centrifugation, remove the supernatant and the silicon oil and prepare the TCA extract for analysis by thin-layer or high-performance liquid chromatography.

[a] A typical assay mixture volume might be 6 ml.
[b] Usually at 50 nCi/ml, though the actual values may vary considerably due to specific labelling of the substrate and the properties of the transport system.
[c] With the time points suggested above, at least two microcentrifuges are required.

If the sedimentation methodology is used to detect the transmembrane solute distribution, that distribution is simply obtained by dividing the (^{14}C-substrate/^{3}H$_2$O) inside ratio by the (^{14}C-substrate/^{3}H$_2$O) outside ratio. If the same methodology is used to determine uptake kinetics, the computation of uptake rates is as for the filtration methodology (see section 2.1.1).

2.1.3 Flow dialysis methodology

During flow dialysis, cells or vesicles are mixed with a radiolabelled solute in the upper of two chambers separated by a dialysis membrane (17) (*Figure 1*). Solute that is not taken up by the cells or vesicles diffuses into the lower chamber, through which buffer flows continuously from a reservoir to a fraction collector. Under steady state conditions, the concentration of solute in each recovered aliquot of buffer reflects that present in the upper chamber at a given time point. The difference between that concentration and the concentration observed in the absence of the cells or vesicles indicates the amount of solute taken up. *Protocol 5* can be used to determine dependence

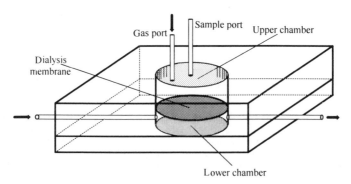

Figure 1. Schematic representation of a flow dialysis apparatus. The volume of the lower chamber can be minimized by employing the capillary design of Feldman (17). Special care must be taken to provide sufficent oxygen by bubbling air through the upper chamber. The high cell densities used in these experiments would otherwise rapidly lead to a state of anaerobiosis.

of the transmembrane distribution of a solute in bacterial membrane vesicles on a particular energy supply.

Protocol 5. Determination of solute uptake by flow dialysis

Equipment and reagents

- Flow dialysis apparatus with a flow adaptor on the chamber inlet port, a peristaltic pump connected to the chamber outlet port, and a water-jacket for temperature control (suitable devices can be quite simply constructed) (17)
- Temperature controlled water-bath
- Fraction collector
- Dialysis membranes
- Two mini-stir bars

- Membrane vesicle suspension, as described in *Protocol 2*
- Solution of cellular energy source (e.g. D-glucose to yield a final concentration of 10 mM)
- Solution of radioisotopically labelled solute with known specific radioactivity (see *Protocol 1, Hint* (e))
- Micropipettors and/or Hamilton syringes
- Assay buffer, bubbled with air

Method

1. Assemble the flow dialysis apparatus with an appropriate dialysis membrane separating upper and lower chamber, water of an appropriate temperature circulating through the water-jacket, and air circulating through the upper chamber. Add an aliquot of assay buffer to the upper chamber and initiate flow of assay buffer through the lower chamber.[a]

2. Add an aliquot of vesicle suspension and the radiolabelled solute to the upper chamber (for *E. coli* vesicles, to attain a final concentration of 1–4 mg protein/ml). Allow perfusion of the lower chamber to occur to establish a baseline rate of solute elution.[b]

3. Initiate uptake by adding a transport energy source to the upper chamber.[c]

4. Samples of the collected fractions are mixed with scintillation fluid and counted in a scintillation counter.

Helpful hint

When choosing a flow dialysis apparatus it is useful to pick one with a big dialysis area, but a low volume lower chamber to minimize the dead volume, thereby greatly enhancing the sensitivity of the system.

[a] Both compartments should contain a mini-stir bar to ensure a homogeneous distribution of radiolabelled solute and cells or vesicles. This is important because of the high cell densities used in this method. The flow of buffer through the lower chamber should be adjusted by a peristaltic pump on the outlet. It is usually approximately 6 ml/min. Care must be taken not to damage the dialysis membrane by applying too high a flow rate.

[b] The initial phase is characterized by an increase in radioactivity of buffer recovered from the lower chamber (the actual fraction size as well as the proportion of each fraction used to test for the amount of label may vary depending on individual experimental parameters).

[c] Energy-dependent uptake of the radiolabelled solute will be indicated by a steady decrease in concentration of radiolabel in the buffer recovered from the lower chamber. Perfusion should continue until a new steady state has been attained.

Flow dialysis is often used to obtain qualitative information regarding the response of particular transporters to various energy sources and inhibitors of energy metabolism. It can be used effectively to estimate transmembrane solute distributions. The quantity of solute accumulated by cells or vesicles during flow dialysis is indicated by the difference in radioactive solute concentration between solvent fractions eluting during uptake (usually in the steady state) and those eluting under control conditions (usually under de-energized conditions). Such values are then combined with estimates of the quantity of cells or vesicles added to the dialysis chamber as described for the filtration methodology (section 2.1.1). To determine the intracellular solute concentration, the cytoplasmic volume must be independently estimated (18).

2.1.4 Determination of the purity of radiolabelled solutes

Unfortunately, purchased radiolabelled compounds are not always chemically pure. Reproducible and accurate uptake data can only be obtained with radiolabelled solutes that are devoid of labelled or unlabelled contaminants. The presence of a labelled contaminant that is taken up by the cells or vesicles under investigation can often be detected through a sudden and dramatic increase in measured uptake values, as such compounds are usually present at high specific radioactivities that have not been reduced by the experimenter. Radiolabelled stock solutions should be tested for purity upon delivery and at intervals during experimentation to exclude the possibility of chemical or radiochemical breakdown. The purity of radiolabelled compounds is readily tested using thin-layer or high-performance liquid chromatography. The compound of interest is mixed with the corresponding unlabelled carrier substance and chromatographed with a suitable stationary/mobile phase system. Special care has to be taken that the solvent system used is the one that gives the best resolution for the separation of the solute from its most likely contaminants. Most manufacturers specify the systems they use for quality control. Those systems are a useful guide, though they are not always optimal for each compound. The chromatogram or fractions are analysed by autoradiography or by scintillation counting, respectively. Unlabelled solutes (and solute analogues, if appropriate) which can be detected chemically serve as reference substances.

2.2 Electrochemical methods

Although the direct electrochemical detection of transmembrane fluxes can provide valuable opportunities for the study of transport processes, it is possible only for black lipid membrane systems, cell, or vesicle systems large enough to accept membrane-penetrating electrodes or systems that can be 'patch-clamped'. Those techniques are beyond the scope of this chapter. Ion fluxes of substantial magnitude can be detected indirectly, however, by monitoring changes in the composition of media in which cells or vesicles are suspended using ion-selective electrodes (19).

pH electrodes are commonly used to detect the transmembrane proton fluxes that accompany organic solute fluxes during solute/proton co-transport. Such measurements are the subject of *Protocol 6*. In order to observe such proton movements, cells or vesicles must be suspended in unbuffered (or very lightly buffered) media, respiration and other metabolic processes that might perturb medium pH must be inhibited (for example, by excluding oxygen, including metabolic inhibitors, or using cells defective for solute metabolism). In addition, a membrane permeant counterion (e.g. SCN^-) must be provided to prevent the development of an inhibitory membrane potential. Proton movements can then be measured as a net pH change in the suspending medium. Other ion-selective electrodes can be used in a similar manner.

In principle, ion coupling stoichiometries can be estimated by comparing ion fluxes determined in this manner with solute fluxes determined independently, under the same conditions, using radiolabelled solutes. In practice, such correlations may be unreliable. Initial rates of solute-coupled ion flux can be difficult to determine since, even in the absence of metabolic effects, simultaneous ion fluxes mediated by other membrane systems may reinforce and counteract the phenomenon under study. While qualitative comparison of the ion and solute fluxes is important, experimenters are advised to determine coupling stoichiometries by comparing solute and ion distributions.

Protocol 6. Detection of proton-linked solute symport in *E. coli*

Equipment and reagents

- Water-jacketed reaction vessel fitted with appropriate electrode, gas, and reagent ports (19)
- Temperature controlled water-bath
- Cell suspension in transport assay buffer (see step 1 or ref. 20)
- Assay buffer (for *E. coli*):150 mM KCl, 5 mM glycylglycine pH 6.6
- Solution of transport solute
- Micropipettors and/or Hamilton syringes
- Nitrogen or argon gas

Method

1. Culture cells in an appropriate medium.[a] Harvest cells by centrifugation and resuspend them in assay buffer supplemented with 1 mM β-mercaptoethanol.

2. Incubate the cell suspension for 1 h at the growth temperature with shaking.[b]

3. Re-centrifuge the cell suspension and resuspend the cells in assay buffer to an optical density (680 nm) of approximately 50 (about 35 mg dry mass/ml).

Protocol 6. *Continued*

4. Insert the pH electrode into the airtight cell and immerse the cell in the temperature controlled water-bath. [c]

5. Displace oxygen from the system by flooding the cell with a light stream of nitrogen or argon gas. The exclusion of oxygen is important at all stages of the measurements.

6. Rinse the measuring cell with distilled water followed by at least three washes with anoxic assay buffer. Finally, just before the cell suspension is introduced, flush the empty cell with argon gas.

7. Add 0.5 ml of the cell suspension to the measuring cell. Immediately, fill the cell with anoxic assay buffer and close the lid. Allow the system to equilibrate for 15–45 min until the pH is stable. [d]

8. Start the measurement by adding the substrate for the transport system in which you are interested. Observe the change in pH on the chart recorder.

Helpful hints

(a) Calibration of the system is vital and is done by measuring the magnitude of the pH change that accompanies addition of known amounts of standard acid or base to the system.

(b) Mutant organisms defective in substrate transport and/or metabolism provide critically important controls in efforts to establish a link between an observed pH change and a transport system of interest.

(c) To measure Na^+ fluxes, suspend cells in tetramethylammonium-*N*-Tris-(hydroxymethyl)-methylglycine (Tricine) (brought to pH 8 with tetramethylammonium hydroxide (TMA-OH)), 10 mM NaCl, and use a Na^+-sensitive electrode. To measure Li^+ fluxes, suspend cells in 100 mM Tris-(hydroxylmethyl)-aminomethane/Morpholinopropane sulfonic acid buffer (pH 7), 100 mM LiCl, and use a Li^+-sensitive electrode.

[a] If possible, use late log or stationary phase cultures for which oxygen is growth limiting to help ensure the depletion of endogenous energy stores.
[b] This helps to remove the growth substrate, and to deplete energy and ATP.
[c] It is critical that the temperature be tightly controlled ($\pm$ 0.1°C) with a water-bath because small fluctuations in temperature may cause pH changes that are similar in size to those being measured. Ideally, the electrode is attached to a strip chart recorder to record pH changes.
[d] If equilibration takes longer, it may indicate that oxygen has not been adequately excluded, that the substrate was insufficiently depleted, or that the measurement cell was insufficiently washed between experiments.

2.3 Spectroscopic methods

Enzymologists routinely exploit spectral differences between enzyme substrates and products (or their derivatives) in designing enzyme assays based on absorption spectroscopy. Unfortunately, absorption spectroscopy has limited application for studies of membrane permeability and transport, since most solute fluxes of interest to experimenters do not involve covalent solute modification. The activity of the lactose permease in whole *E. coli* cells can be estimated using substrate analogue *o*-nitrophenyl galactoside (ONPG), however, since post-flux hydrolysis of ONPG by β-galactosidase releases the chromophore *o*-nitrophenol. ONPG is membrane permeant and hence equilibrates throughout the reaction mixture, allowing spectroscopic monitoring of transporter activity. Identification of similar possibilities for other transport systems may greatly simplify their detection and analysis.

As Mitchell indicated in formulating the chemiosmotic hypothesis, transmembrane solute fluxes alter transmembrane osmolality gradients. Since biological membranes are highly water permeable, water spontaneously redistributes in response to net solute fluxes. Such osmotic phenomena may be expected to cause swelling or shrinkage of the cellular cytoplasm or vesicle lumen, causing changes in particle size and/or shape detectable through integrated or dynamic light scattering. Such volume changes are, of course, modulated by cell walls. Light scattering techniques can, in principle, be used to detect transmembrane solute fluxes, particularly in (proteo)liposome and membrane vesicle systems (21, 22). Recent work has revealed diverse osmoregulatory mechanisms through which cells resist osmotically-induced volume changes. Experimenters using light scattering to detect volume changes in intact cells or vesicles must therefore be aware that experimentally imposed solute fluxes may trigger countervailing osmoregulatory responses.

Light scattering can be used to detect changes in particle size, shape, and structure. Simple nephelometry, accomplished with instruments designed for absorption spectroscopy, can be used to obtain qualitative information about such changes. To fully define relationships among particle size, shape, and structure, measurements must be made with instruments that record the angular dependence of the intensity of scattered laser light (integrated light scattering). Dynamic light scattering, which also requires specialized instrumentation, is sensitive to the diffusive motions of particles and is therefore particularly sensitive to particle size and shape (22, 23). Since sophisticated light scattering instrumentation is not available to most investigators, no protocol for those measurements is provided here. However, they can provide a particularly effective method for the detection and characterization of passive transmembrane fluxes.

3. Transporter classes

Using the methodologies described above, experimenters continue to identify diverse transporters present in all living cells (*Table 2*). In combination with other approaches such experiments are leading to the formulation of detailed transport mechanisms. For bacterial systems, current research foci include

Table 2. Classification of transporters by structure and energy supply

Transporter type	Transporter structure	Energy supply	Examples
Facilitator	Membrane integral polypeptide	$\Delta\tilde{\mu}_{\text{Substrate}}$	Glycerol facilitator (*E. coli*) Glucose facilitator GLUT1 (mammalian erythrocytes)
Ion-solute co-transporter	Membrane integral polypeptide	$\Delta\tilde{\mu}_1$	Lactose permease (*E. coli*) Glucose transporter (mammalian cells) Na^+/H^+ antiporter (*E. coli*)
Traffic ATPase	Multiple extramembrane (periplasmic or cell surface), membrane peripheral, and membrane integral polypeptides or polypeptide domains	ATP	Maltose transporter (*E. coli*) MDR (multidrug efflux pump of mammalian cells)
Group translocator	Multiple extramembrane (cytoplasmic), membrane peripheral, and membrane integral polypeptides or polypeptide domains	PEP	Sugar–PEP phosphotransferases (*E. coli*)

the extension of transport studies to diverse organisms and the study of export systems for small and large molecules. For systems present in higher eukaryotes, the implementation of molecular biological approaches is now facilitating the resolution of transporters with overlapping specificities, and supporting analyses of their tissue distribution, developmentally regulated expression, and physiological function.

Acknowledgements

This work was supported by the Natural Sciences and Engineering Research Council of Canada and the Deutsche Forschungs Gemeinschaft. We are grateful to F. Ross Hallett (University of Guelph) for his comments on the detection of membrane permeability and transport using light scattering techniques, and to Drs E. Padan (The Hebrew University of Jerusalem) and E. Bakker (University of Osnabrück) for their comments on the manuscript.

References

1. Mitchell, P. (1961). *Nature*, **191**, 144.
2. Tortorella, D., Ulbrandt, N. D., and London, E. (1993). *Biochemistry*, **32**, 9181.

3. Grothe, S., Krogsrud, R. L., McClellan, D. J., Milner, J. L., and Wood, J. M. (1986). *J. Bacteriol.*, **166**, 253.
4. Wood, J. M. (1975). *J. Biol. Chem.*, **250**, 4477.
5. Krupka, R. M. (1992). *Biochim. Biophys. Acta*, **1110**, 1.
6. Sanders, D., Hansen, U.-P., Gradmann, D., and Slayman, C. L. (1984). *J. Membr. Biol.*, **77**, 123.
7. Kaczorowski, G. J., Robertson, D. E., and Kaback, H. R. (1979). *Biochemistry*, **18**, 3697.
8. Kaczorowski, G. J. and Kaback, H. R. (1979). *Biochemistry*, **18**, 3691.
9. Walmsley, A. R., Petro, K. R., and Henderson, P. J. F. (1993). *Eur. J. Biochem.*, **215**, 43.
10. Rottenberg, H. (1976). *FEBS Lett.*, **66**, 159.
11. Chen, C.-C., Tsuchiya, T., Yamane, Y., Wood, J. M., and Wilson, T. H. (1985). *J. Membr. Biol.*, **84**, 157.
12. Bakker, E. P. (1990). *FEMS Microbiol. Rev.*, **75**, 319.
13. Kaback, H. R. (1971). In *Methods in enzymology* (ed. W. B. Jakoby), Vol. 22, pp. 99–129. Academic Press, London.
14. Dupont, Y. (1984). *Anal. Biochem.*, **142**, 504.
15. Heller, K. B., Lin, E. C. C., and Wilson, T. H. (1980). *J. Bacteriol.*, **144**, 274.
16. Voegele, R. T., Sweet, G. D., and Boos, W. (1993). *J. Bacteriol.*, **175**, 1087.
17. Feldmann, K. (1978). *Anal. Biochem.*, **86**, 555.
18. Stock, J. B., Rauch, B., and Roseman, S. (1977). *J. Biol. Chem.*, **252**, 7850.
19. Henderson, P. J. F. and MacPherson, A. J. S. (1986). In *Methods in enzymology* (ed. S. Fleischer and B. Fleischer), Vol. 125, pp. 387–429. Academic Press, London.
20. Zilberstein, D., Schuldiner, S., and Padan, E. (1979). *Biochemistry*, **18**, 669.
21. de Gier, J. (1993). *Chem. Phys. Lipids*, **64**, 187.
22. Hallett, F. R., Watton, J., and Krygsman, P. (1991). *Biophys. J.*, **59**, 357.
23. Ertel, A., Marangoni, A. G., Marsh, J., Hallet, F. R., and Wood, J. M. (1993). *Biophys. J.*, **64**, 426.

3

Measurement of mitochondrial protonmotive force

MARTIN D. BRAND

1. Introduction

Protonmotive force is the central intermediate in bioenergetics. Mitochondria normally produce it by pumping protons from the mitochondrial matrix across the inner membrane during electron transport. Protonmotive force drives the synthesis of ATP by the ATP synthase as well as several other important bioenergetic reactions such as ion transport, transhydrogenation, and proton leak-mediated heat production.

The electrochemical potential difference between protons in the aqueous phases on different sides of any membrane, $\Delta\mu_{H^+}$, has units of kJ/mol. In bioenergetics, by analogy with electromotive force, $\Delta\mu_{H^+}$ is conventionally expressed as protonmotive force, in mV. Protonmotive force is usually abbreviated to pmf or Δp and is equal to $-\Delta\mu_{H^+}/F$, where F is the Faraday constant. For a fuller description see ref. 1.

Like any other electrochemical potential, Δp has two components; an electrical term and a concentration term. These are related by the equation:

$$\Delta p = \Delta\psi - z\Delta pH$$

where $\Delta\psi$ is the electrical potential difference or membrane potential (cytoplasm − matrix), z is $2.303RT/F$, with values of 59 mV at 25 °C and 61.5 mV at 37 °C, and ΔpH is the difference in pH (cytoplasm − matrix). Note that the sign convention used in bioenergetics is opposite to that used in electrophysiology, where the membrane potential across the plasma membrane is conventionally defined as the electrical potential difference (cytoplasm − exterior). This means that a positive outside potential is given a positive sign in bioenergetics and a negative sign in electrophysiology, which can lead to confusion in cellular bioenergetics where both potentials are measured. Typically, Δp in mitochondria might have a value of 180 mV, consisting of 150 mV of $\Delta\psi$ (positive outside) and 30 mV of $-z\Delta pH$ (0.5 pH unit, acid outside). So, when measuring Δp, one must always take into account both membrane potential and pH gradient.

What determines the relative values of $\Delta\psi$ and $-z\Delta pH$ in mitochondria? Because the buffering power of the mitochondrial matrix and the extra-mitochondrial medium is quite high, the pH changes caused by translocation of a few protons across the membrane are quite small. Because the capacitance of the inner membrane is large, translocation of the same few protons will cause large changes in membrane potential. Thus $\Delta\psi$ is normally far greater than $-z\Delta pH$. However, any secondary movements of charged species across the membrane (cation uptake or anion efflux) will reduce $\Delta\psi$. This in turn will cause the electron transport chain to pump more protons, so returning Δp to near its original value but increasing $-z\Delta pH$ at the expense of $\Delta\psi$. This occurs if there is calcium uptake on the endogenous calcium uniporter, or potassium uptake catalysed by added uniport ionophores such as valino-mycin. Conversely, any secondary influx of protons by electroneutral path-ways (cation/proton antiport or anion/proton symport) will reduce $-z\Delta pH$ and lead to a compensatory increase in $\Delta\psi$ through additional proton pump-ing. This occurs if there is phosphate or acetate uptake with concomitant proton influx, or potassium efflux catalysed by added potassium/proton anti-porters such as nigericin. Thus the relative values of the two components of Δp are determined by the pH buffering power in the system, by the capacitance of the inner membrane, by the ion composition of the extra-mitochondrial medium, and by the presence of endogenous or exogenous carriers and ionophores of different types. For more details see ref. 1.

2. Principles of the measurement of Δp

All quantitative methods to measure Δp and its components rely directly or indirectly on measurement of the equilibrium distribution of probe molecules across the mitochondrial inner membrane. It is important that the concentration of probe used is very low, otherwise its movement across the membrane will have a significant effect on the gradient that is being measured.

2.1 Measurement of $\Delta\psi$

Probes to measure $\Delta\psi$ are charged. They are usually positively charged to ensure that they are accumulated into the matrix. Provided that there is no active transport, a monovalent cation (C^+) will eventually come to the equilibrium described by the Nernst equation:

$$\Delta\psi = (RT/F) \ln ([C^+]_{in}/[C^+]_{out})$$
$$\Delta\psi = 61.5 \log ([C^+]_{in}/[C^+]_{out}) \text{ at } 37°C.$$

The accumulation of the probe molecules is monitored by following some appropriate change in absorbance or fluorescence, by the use of radiolabels, or by measurement of the external concentration with ion-selective electrodes.

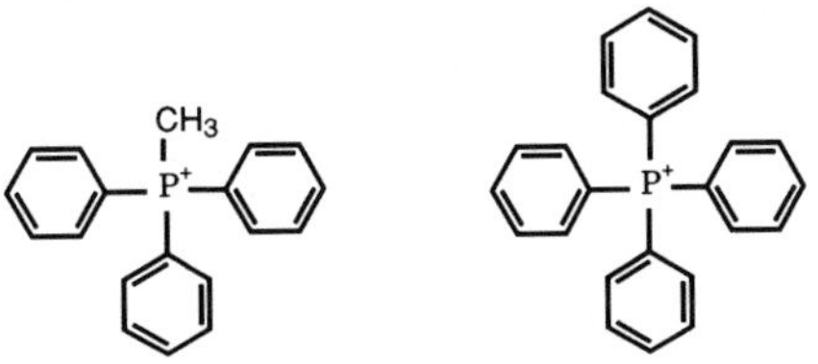

TPMP   TPP

Figure 1. The structures of methyltriphenylphosphonium, TPMP$^+$, and tetraphenylphosphonium, TPP$^+$.

Dyes such as safranine and rhodamine change their absorbance when they are accumulated, due to stacking effects at the membrane surface or in the matrix, and so give optical signals that can be used qualitatively or can be calibrated against known potentials set or measured by exploiting the Nernst relationship. These probes will not be considered further in this chapter.

There are two commonly used types of radiolabelled probe. ^{86}Rb$^+$ equilibrates across membranes according to the Nernst equation in the presence of valinomycin. Amphipathic (lipophilic) organic ions such as methyltriphenylphosphonium (triphenylmethylphosphonium, TPMP$^+$)or tetraphenylphosphonium (TPP$^+$) (*Figure 1*) do the same, but do not require the addition of valinomycin. They are commercially available labelled either with ^{14}C or with ^{3}H.

Distribution of cations can also be monitored with electrodes that report the extramitochondrial concentration. From the change in external concentration and the volume of the mitochondrial matrix it is possible to calculate the intramitochondrial concentration, and so the value of $\Delta\psi$. Potassium distribution in the presence of valinomycin, or TPMP or TPP distribution are often measured in this way.

2.2 Measurement of ΔpH

Probes to measure ΔpH are lipid soluble weak acids and bases. Provided that there is no active transport, a weak acid (AH) that crosses the membrane only when protonated will eventually come to equilibrium according to ΔpH. At equilibrium:

$$\Delta pH = pH_{out} - pH_{in}$$
$$= \log ([H^+]_{in}/[H^+]_{out})$$
$$= -\log ([A^-]_{in}/[A^-]_{out}).$$

If the pK of the acid is sufficiently below the pH on either side of the membrane so that most of the acid is ionized to A$^-$ then this approximates to:

$$\Delta pH = -\log ([A]_{in}/[A]_{out}).$$

The accumulation of the radiolabelled weak acid can be measured in the same way as the accumulation of the radiolabelled $\Delta\psi$ probes.

It is also possible to use pH indicators to measure ΔpH, if the signals from the internal and external phases can be distinguished. One way to achieve this is to use a permeant indicator such as neutral red, but to buffer the external phase so strongly that the only signal that changes comes from the internal phase. Nuclear magnetic resonance signals can also be used. In each case calibration of the signal is required.

3. Measurement of Δp in mitochondria

To measure Δp, both $\Delta\psi$ and ΔpH should be measured. It is sometimes convenient to avoid the need to measure ΔpH by clamping it to zero, so that the whole of Δp is expressed as $\Delta\psi$. This can be done by adding nigericin to mitochondria in a medium that has a relatively high potassium concentration (120 mM) so that there is no potassium gradient across the membrane. The nigericin catalyses potassium/proton antiport and sets the ΔpH to zero, and the electron transport chain ensures that Δp is maintained near its original value, but is expressed entirely as $\Delta\psi$. This trick is inappropriate if valinomycin is to be added in the experiment, as the combination of potassium uniport by the valinomycin and potassium/proton exchange by the nigericin will act as a potent uncoupler, so dissipating Δp. To some extent this disadvantage can be overcome by adding high concentrations of phosphate or acetate instead of nigericin; electroneutral uptake of the anions will tend to consume ΔpH and lower its value. *Protocols 1 to 10* are designed to measure Δp at 37 °C, but they also work well down to at least 20 °C.

3.1 Advantages and disadvantages of radioactive and ion-sensitive electrode methods

Radioactive methods are:

(a) Easy to use in most laboratories since no specialized equipment other than a scintillation counter is required.

(b) Suitable for large numbers of samples in parallel, when results are not needed in real time.

(c) More sensitive than electrode methods at low values of $\Delta\psi$ or ΔpH, since it is the amount of probe in the mitochondria that is measured.

(d) Particularly useful (with ^{86}Rb) if it is important to avoid possible errors due to binding of lipophilic cations.

(e) Sometimes constrained by the relatively short half-life of isotopes such as [86]Rb (half-life 19 days).

(f) More labour-intensive, and more prone to operator error because of the relatively large amount of sample processing that is involved.

(g) Subject to secondary problems due to the need for proper authorization for radioactive work, to low level exposure of the experimenter to radioactivity, and to safe waste disposal.

(h) Relatively expensive in radiochemicals, scintillant, and disposable plastic-ware.

(i) The best way to measure matrix volumes, ΔpH, and lipophilic cation binding corrections.

TPMP and TPP electrode methods are:

(a) More specialized and awkward to set up from scratch, because the electrodes have to be constructed, but very easy to use once in place.

(b) Suitable for small numbers of experiments. Results are obtained in real time, allowing quick assessment of time courses and effects of added reagents. This allows modification of experiments in flight.

(c) Particularly good for titrations of respiration rate and potential, where both are measured simultaneously through a succession of steady states in a single experiment.

(d) Very sensitive to small changes in potential at high potentials, but insensitive and prone to interpretational problems at low potentials since it is the concentration of probe in the external phase that is measured.

(e) Sometimes prone to electrode 'drift', requiring more experience to trouble-shoot.

(f) Prone to artefacts caused by additions of hydrophobic compounds that interact with the electrodes (this can be minimized by calibrating *after* addition of the offending compound).

(g) Very inexpensive.

3.2 Advantages and disadvantages of different probes

Accumulation of $^{86}Rb^+$ or K^+ is:

(a) Dependent on addition of valinomycin, which makes it inappropriate for many experiments since the potassium gradient will tend to clamp $\Delta\psi$, and therefore unsuitable in the presence of nigericin to clamp ΔpH because of the uncoupling effect of the two ionophores together.

(b) Not subject to significant artefacts due to probe binding.

(c) Quick to come to equilibrium, typically in a few seconds (depending on the valinomycin concentration).

Accumulation of TPMP or TPP is:

(a) Independent of added ionophores, so less prone to disturb the system being measured, although both TPMP and TPP inhibit respiration if used at greater than 5–10 μM.

(b) Subject to significant artefacts due to probe binding. These must be corrected for if results are to be expressed quantitatively. TPMP is less lipid soluble than TPP, so binding is less of a problem.

(c) Relatively slow to equilibrate. TPP is faster than TPMP, typically TPP equilibrates within 30 sec compared to 2 min for TPMP in isolated mitochondria.

3.3 Measurement of mitochondrial volume

Whether radioactive or ion-selective electrode methods are used it is necessary to measure (or assume!) the mitochondrial matrix volume so that the concentration of the probe in the matrix can be calculated from the amount of probe in this compartment (*Protocols 1* and *2*).

3.3.1 Principle of the method

If mitochondria are incubated with a radiolabelled probe such as ^3H_2O and then sedimented, the accessible volume of the pellet can be calculated from the specific activity of the probe in the supernatant and the total radioactivity in the pellet. The difference in accessible volume ('pellet space') for permeant probes like ^3H_2O and probes like $[^{14}C]$sucrose that do not cross the inner membrane reports the volume of the matrix. This volume includes the volume occupied by exchangeable water of hydration but not that occupied by other molecules such as proteins or lipids.

Protocol 1. Measurement of mitochondrial volume (2–4)

Equipment and reagents

- Water-bath
- Automatic pipettes
- Mitochondria (50 mg protein/ml)
- Incubation medium: 120 mM KCl, 5 mM Hepes, 1 mM EGTA, 5 μM rotenone, brought to pH 7.2 with KOH, and maintained at 37°C
- ^3H_2O (100 μCi/ml)
- $[^{14}C]$sucrose (10 μCi/ml)

Method

1. Observe the normal precautions for working with low levels of radioactivity.

2. Incubate 40 μl mitochondria (2 mg protein) in 1 ml of medium containing 10 μl (1 μCi) of ^3H_2O, 10 μl (0.1 μCi) of $[^{14}C]$sucrose, plus any other desired additions, for 2 min in a 2 ml plastic centrifuge tube maintained at 37°C in a water-bath.[a]

3. Sediment the mitochondria and measure pellet spaces as described in *Protocol 2*.

4. Calculate matrix volume in μl/mg protein as (^3H_2O space−$[^{14}C]$sucrose space)/mg protein added.

---

5. Average the values from at least six tubes for accurate results.

[a] Incubation time needs to be long enough to allow probe equilibration (probably within the mixing time) but not so long that significant sucrose uptake or metabolism occurs (little uptake within 20 min). If the mitochondrial suspension or the medium do not contain sucrose, add 100 μM sucrose as carrier.

Protocol 2. Determination of pellet spaces (4)

Equipment and reagents

- Bench microcentrifuge
- Vortex mixer
- Scintillation counter
- Automatic pipettes
- Centrifuge tube clippers

- 20% (v/v) Triton X-100
- Scintillation fluid (e.g. LKB Optiphase Hisafe II)
- Paper tissues

Method

1. Sediment the mitochondria by centrifugation at full speed (about 12 000 g) for 2 min in a bench microcentrifuge. [a]

2. For each pellet carry out steps 3 to 6 before processing the next pellet so that volatile radiolabels like ^3H_2O are not lost.

3. Take 500 μl of supernatant into a 5 ml scintillation mini-vial and add 3.5 ml of scintillant.

4. Decant the remainder of the supernatant.

5. Remove supernatant that adheres to the pellet by capillary action using a twisted paper tissue. Make sure that the tube walls and lid are free of residual supernatant.

6. Add 40 μl of Triton X-100 to the pellet. Pellets can be left until the end of the experiment or even overnight at this stage if necessary.

7. Resuspend the pellet by vigorous vortex mixing.

8. Cut the base of the centrifuge tube containing the whole resuspended pellet into a scintillation vial. [b] Add 3.5 ml of scintillant. For greater accuracy add sufficient distilled water (about 200 μl) to equalize the subsequent quenching of scintillation vials between pellet and supernatant samples.

9. Vortex all scintillation vials to mix well, ensuring that the dissolved pellets are mixed out of the tube bases.

10. Count the radioactivity in the vials using a two-channel scintillation counter with appropriate background, cross-over, and quench corrections.

Protocol 2. *Continued*

11. Calculate pellet spaces in μl as [pellet d.p.m. × supernatant volume (=500 μl)]/supernatant d.p.m.

> [a] Most of the mitcochondria will pellet within 15 sec; the remainder of the spin compacts the pellet for the next step.
> [b] Dog toe-nail clippers sold in pet shops are ideal for this.

3.3.2 Trouble-shooting mitochondrial volume measurements

Most novices find that their initial measurements of mitochondrial volume are very variable, and frequently give negative values. Results tend to improve after one or two repetitions. Values for energized rat liver mitochondria should be about 0.5 to 0.8 μl/mg protein; for de-energized mitochondria the values can rise above 1.0 μl/mg protein, depending on the conditions.

(a) A plot of volume against protein added tends not to pass through the origin, so always use a relatively high protein concentration of mitochondrial protein to minimize artefacts due to probe binding to tubes, evaporation of ^3H_2O, etc. Substitution of $[^3H]$ urea for ^3H_2O can bypass evaporation problems.

(b) Because the determination of volume requires measurement of the difference between two relatively error-prone spaces, noise is always quite high. Do sufficient replicates (at least four, preferably six) to allow good averages to be obtained.

(c) Work quickly to minimize probe evaporation. Significant evaporation of ^3H_2O from the tube walls or the pellet surface can occur if damp surfaces are left exposed to air for too long, and this will reduce the ^3H_2O pellet space but not the $[^{14}C]$sucrose pellet space.

(d) Be sure to have five or ten times more ^3H counts than ^{14}C counts so that any consistent errors in cross-over correction during scintillation counting do not have much impact.

(e) Inadequate resuspension of pellets or mixing of scintillation vials can lead to unequal quenching and variability of results.

(f) A poorly calibrated scintillation counter can affect the results with different isotopes to different extents.

(g) Bacterial or other contamination of $[^{14}C]$sucrose stocks can lead to overestimation of pellet sucrose spaces.

(h) Some authors prefer to use $[^{14}C]$mannitol instead of $[^{14}C]$sucrose.

(i) Some authors prefer to pellet through oil (see e.g. *Protocol 12*) for greater accuracy. However, we find that accuracy is sufficient and the experiments are much easier and cleaner if the mitochondria are simply pelleted, *Protocol 2*.

Despite these potential problems, the effect of errors in the measurement of mitochondrial volumes is often not very great, since a twofold error in the volume causes only an 18 mV error in the calculated value of $\Delta\psi$, or ΔpH, since 61.5 log 2 = 18.5 mV.

3.4 Measurement of probe accumulation using radioactivity

If the apparent matrix space available to a probe molecule is divided by the true matrix space, then the extent to which the probe is accumulated can be calculated. The accumulation ratio is $[\text{probe}]_{in}/[\text{probe}]_{out}$. $\Delta\psi$ and ΔpH can be calculated from accumulation ratios of suitable probes.

3.4.1 Determination of probe accumulation (2–5)

See *Protocols 3–5*. In all cases make sure that mitochondrial suspensions do not become anaerobic or cooled before centrifugation as this will change the values that are being determined.

Protocol 3. Measurement of ΔpH (2–5)

Equipment and reagents

- Water-bath
- Automatic pipettes
- Mitochondria (50 mg protein/ml)
- Incubation medium: 120 mM KCl, 5 mM Hepes, 1 mM EGTA, 5 μM rotenone, brought to pH 7.2 with KOH, and maintained at 37°C

- [^{3}H]acetate (100 μCi/ml)
- [^{14}C]sucrose (10 μCi/ml)
- Acetate (10 mM)
- Sucrose (10 mM)

Method

1. Incubate 40 μl mitochondria (2 mg protein) in 1 ml of medium containing 10 μl (1 μCi) of [^{3}H]acetate[a] and 10 μl (0.1 μCi) of [^{14}C]sucrose for 2 min in a 2 ml plastic centrifuge tube in a water-bath.[b] Include 10 μl of acetate (100 μM) and 10 μl of sucrose (100 μM) as carriers if required.

2. Determine matrix volume in parallel as described in *Protocol 1*.

3. Sediment the mitochondria and measure pellet spaces as described in *Protocol 2*.

4. Calculate acetate accumulation ratio as ([^{3}H]acetate space – [^{14}C]sucrose space)/matrix volume.

5. To measure a reverse pH gradient (acid matrix), use 1 μCi of [^{3}H]sucrose and 0.1 μCi of [^{14}C]methylamine, and calculate methylamine accumulation ratio as ([^{14}C]methylamine space – [^{3}H]sucrose space)/matrix volume. Include carrier methylamine (20 μM) and sucrose (100 μM) if required (see ref. 3).

Protocol 3. *Continued*

6. Calculate ΔpH as $-\log$ (acetate accumulation ratio) or as $+ \log$ (methyl-amine accumulation ratio). Multiply by $- 2.303RT/F$ (-61.5) to calculate $-z\Delta pH$. [c]

7. Average the values from at least three tubes for accurate results.

[a] Some authors prefer to use [³H]lactate or [³H]dimethyloxazolidinedione (DMO) to reduce metabolism of the label, but we have not found this to be an improvement.

[b] Incubation time needs to be long enough to allow probe equilibration (about 1–2 min) but not so long that significant sucrose uptake or acetate metabolism occurs (negligible within 2 min).

[c] For example, tenfold accumulation of acetate gives ΔpH of -1.0 pH units and $-z\Delta pH$ of 61.5 mV at 37°C.

Protocol 4. Measurement of $\Delta\psi$ using ⁸⁶Rb (5)

Equipment and reagents

- Water-bath
- Automatic pipettes
- Mitochondria (50 mg protein/ml)
- Incubation medium: e.g. 120 mM LiCl, 5 mM succinic acid, 5 mM Hepes, 1 mM EGTA, 5 μM rotenone, all brought to pH 7.2 with LiOH, and maintained at 37°C

- ⁸⁶RbCl (2.5 μCi/ml)
- [³H]sucrose (100 μCi/ml)
- Valinomycin (0.4 mg/ml EtOH)
- Sucrose (10 mM)

Method

1. Incubate 40 μl mitochondria (2 mg protein) in 1 ml of medium containing 10 μl (0.025 μCi) of ⁸⁶RbCl (and 10 μl (1 μCi) of [³H]sucrose for greatest accuracy) for 2 min in a 2 ml plastic centrifuge tube in a water-bath, in the presence of 1 μl (0.4 μg) valinomycin/ml. [a] Add 10 μl of carrier sucrose (100 μM) if required.

2. Determine matrix volume in parallel as described in *Protocol 1*.

3. Sediment the mitochondria and measure pellet spaces as described in *Protocol 2*.

4. Calculate rubidium accumulation ratio as (⁸⁶Rb space– [³H]sucrose space)/matrix volume. For high values of $\Delta\psi$ (greater than 150 mV), sucrose space is negligible and may be assumed or ignored.

5. Calculate $\Delta\psi$ as 61.5 log (Rb accumulation ratio) (at 37°C). [b]

6. Average the values from at least three tubes for accurate results.

[a] The potassium concentration in the medium will have a major impact on the value of $\Delta\psi$; use sucrose-based or Li-based media for high values (for example, 120 mM LiCl, 5 mM succinic acid, 5 mM Hepes, 1 mM EGTA, 5 μM rotenone, all brought to pH 7.2 with LiOH). Incubation time needs to be long enough to allow probe equilibration (about 30 sec) but not so long that significant sucrose uptake or metabolism occurs (negligible within 2 min).

[b] For example, 100-fold accumulation of Rb corresponds to $\Delta\psi = 61.5 \log 100 = 123$ mV.

Protocol 5. Measurement of $\Delta\psi$ using [^{3}H]TPMP (2–4)
[^{3}H]TPP can be used in the same way

Equipment and reagents

- Water-bath
- Automatic pipettes
- Mitochondria (50 mg protein/ml)
- Incubation medium: 120 mM KCl, 5 mM Hepes, 1 mM EGTA, 5 μM rotenone, brought to pH 7.2 with KOH, and maintained at 37°C

- [^{14}C] sucrose (10 μCi/ml)
- [^{3}H]TPMP (100 μCi/ml)
- Sucrose (10 mM)
- TPMP (100 μM)

Method

1. Incubate 40 μl mitochondria (2 mg protein) in 1 ml of medium containing 10 μl (1 μCi) of [^{3}H]TPMP, 10μl (1 μM) unlabelled TPMP, and 10 μl (0.1 μCi) of [^{14}C]sucrose for 2 min in a 2 ml plastic centrifuge tube in a water-bath.[a] Add 10 μl of carrier sucrose (100 μM) if required.

2. Determine matrix volume in parallel as described in *Protocol 1*.

3. Sediment the mitochondria and measure pellet spaces as described in *Protocol 2*.

4. Calculate TPMP accumulation ratio as ([^{3}H]TPMP space − [^{14}C]sucrose space)/matrix volume. For high values of $\Delta\psi$ (greater than 120 to 150 mV), sucrose space is negligible and may be assumed or ignored.

5. Calculate $\Delta\psi$ as 61.5 log (TPMP accumulation ratio × TPMP binding correction A) (at 37°C).[b] TPMP binding correction A is determined as described in *Protocol 6*.

6. Alternatively, measure TPMP binding correction B as described in *Protocol 7*. In this case, omit step 2 and calculate $\Delta\psi$ as 61.5 log {([^{3}H]TPMP space − [^{14}C]sucrose space) × TPMP binding correction B} (at 37°C).[c]

7. Average the values from at least three tubes for accurate results.

[a] Incubation time needs to be long enough to allow probe equilibration (1–2 min) but not so long that significant sucrose uptake or metabolism occurs (negligible within 2 min).
[b] For example, a TPMP accumulation ratio of 1000 might correspond to $\Delta\psi$ of 61.5 log (1000×0.32) = 154 mV.
[c] For example, a TPMP pellet space of 600 μl/mg might correspond to $\Delta\psi$ = 61.5 log {(600−4) × 0.55} = 155 mV.

3.4.2 Measurement of probe binding

Since probes such as TPMP and TPP are sufficiently hydrophobic to be able to cross membranes, they are also prone to bind. This binding can be considerable and must be corrected for if quantitative results are required. It is best to use an empirical correction, calibrating the signal against a probe that does not bind (such as ^{86}Rb), rather than a theoretical correction that may not be accurate. For TPMP (and TPP) there are two different correction protocols, A and B. Correction A (*Protocol 6*) (2, 6) adjusts the accumulation ratio for TPMP to the accumulation ratio for ^{86}Rb measured over a range of membrane potentials, and ignores any effect of matrix volume on the relative binding of TPMP. It is conceptually simpler and slightly easier to do, but it is less accurate if there is a significant volume dependence of TPMP binding and it does not eliminate the normal requirement that matrix volume is measured in paralled in all experiments. Correction B (*Protocol 7*) (7) adjusts the accumulation ratio for TPMP to the accumulation ratio for ^{86}Rb measured over a range of matrix volumes, and ignores any effect of $\Delta\psi$ on the relative binding of TPMP. It is conceptually more subtle and slightly harder to do, but it is more accurate and has the important advantage that it allows measurement of $\Delta\psi$ by either radioactive or electrode methods without the requirement for routine parallel measurements of matrix volume. One unavoidable disadvantage of binding correction B is that it is necessarily carried out in a different medium from the usual experimental one, requiring the assumption that the difference in media is unimportant for TPMP binding.

Corrections for binding of acetate or methylamine are less usual, since these probes tend to bind less. However, the absolute values of $-z\Delta pH$ are smaller and so errors introduced by binding corrections (which are equivalent to subtracting a constant number of millivolts from the measured values) are correspondingly more important. Probe distribution can be adjusted to ^{86}Rb distribution in the presence of nigericin (3).

Protocol 6. Measurement of correction factor for TPMP binding, method A

Equipment and reagents

- Water-bath
- Automatic pipettes
- Mitochondria (50 mg protein/ml)
- Incubation medium: 200 mM sucrose, 5 mM Hepes, 5 mM LiCl, 1 mM EGTA, 5 µM rotenone, 1 µM TPMP, plus either 0, 0.2, 1.0, or 5.0 mM KCl, neutralized to the required pH (e.g. 7.2) with LiOH or tetramethylammonium hydroxide (TMAOH), and maintained at 37°C

- ^{86}RbCl (2.5 µCi/ml)
- [^{3}H]TPMP (10 µCi/ml)
- ^{3}H$_2$O (100 µCi/ml)
- [^{14}C]sucrose (10 µCi/ml)
- Valinomycin (0.4 mg/ml EtOH)
- Succinate (500 mM, brought to pH 7 with LiOH or TMAOH)

Method

1. Incubate 40 µl of mitochondria (2 mg protein) in triplicate in 1 ml of medium.

2. Add 1 µl (0.4 µg) valinomycin/ml, plus either 10 µl (0.025 µCi) of ^{86}RbCl and 10 µl (0.1 µCi) of [^{3}H]TPMP, or 10 µl (1 µCi) of ^{3}H$_2$O and 10 µl (0.1 µCi) of [^{14}C]sucrose.

3. Add 10 μl (5 mM) succinate (Li or TMA salt).

4. After 2 min sediment the mitochondria and determine pellet spaces as described in *Protocol 2*.

5. Calculate matrix volumes as (^{3}H$_2$O space $-$ [^{14}C]sucrose space), and Rb and TPMP accumulation ratios as (^{86}Rb space $-$[^{14}C]sucrose space)/matrix volume and ([^{3}H]TPMP space $-$ [^{14}C]sucrose space)/ matrix volume.

6. Plot Rb accumulation ratio against TPMP accumulation ratio. The slope of this line is the potential-dependent TPMP binding correction (fraction of TPMP in the matrix that is free) determined by method A.[a]

7. Average at least three experiments for accurate results.

[a] Typical values range from 0.3 to 0.4, indicating that 70% to 60% of the TPMP in the matrix is bound. In principle a correction for any intercept (the potential-independent binding correction) should also be applied, but such a correction is trivial except at very low values of $\Delta\psi$ and is in any case not very reliable. The correction factor is a dimensionless quantity.

Protocol 7. Measurement of correction factor for TPMP binding, method B

Equipment and reagents

- Water-bath
- Automatic pipettes
- Mitochondria (50 mg protein/ml)
- Incubation medium: 5 mM Hepes, 1 mM EGTA, 5 μM rotenone, 1 μM TPMP, plus either 140, 180, 240, or 300 mM sucrose as osmotic support, neutralized to the required pH (e.g. 7.2) with LiOH or TMAOH, and maintained at 37°C

- ^{86}RbCl (2.5 μCi/ml)
- [^{3}H]TPMP (10 μCi/ml)
- ^{3}H$_2$O (100 μCi/ml)
- [^{14}C]sucrose (10 μCi/ml)
- Valinomycin (0.4 mg/ml EtOH)
- Succinate (500 mM, brought to pH 7 with LiOH or TMAOH)

Method

1. Incubate mitochondria (2 mg protein) in triplicate in 1 ml of medium.

2. Add 1 μl (0.4 μg) valinomycin/ml, plus either 10 μl (0.025 μCi) of ^{86}RbCl and 10 μl (0.1 μCi) of [^{3}H]TPMP, or 10 μl (1 μCi) of ^{3}H$_2$O and 10 μl (0.1 μCi) of [^{14}C]sucrose.

3. Add 10 μl (5 mM) succinate (LiOH or TMAOH salt).

4. After 2 min sediment the mitochondria and determine pellet spaces as described in *Protocol 2*.

5. Calculate matrix volume as (^{3}H$_2$O space $-$ [^{14}C]sucrose space)/mg protein added and Rb and TPMP accumulation ratios as (^{86}Rb space $-$ [^{14}C]sucrose space)/matrix volume and ([^{3}H]TPMP space $-$ [^{14}C] sucrose space)/matrix volume.

Protocol 7. *Continued*

6. Divide Rb accumulation ratio by TPMP accumulation ratio and plot it against matrix volume. Draw a line to pass through the origin. If it clearly does not do so, use method A instead. The slope of the line is the volume-dependent TPMP binding correction determined by method B. It has units of $(\mu l/mg\ protein)^{-1}$ and represents the fraction of TPMP in the matrix that is free per $\mu l/mg$.[a]

7. Average at least three experiments for accurate results.

---

[a] A typical value would be between 0.4 and 0.7 $(\mu l/mg\ protein)^{-1}$.

3.5 Measurement of probe accumulation using ion-sensitive electrodes

Electrodes non-selectively sensitive to a range of hydrophobic cations such as TPMP and TPP are quick, easy and cheap to construct (*Protocol 8*), and give reliable results once the art of using them is mastered. Essentially they consist of a polyvinylchloride membrane impregnated with tetraphenylboron, which provides the hydrophobic negative charge to interact with the cation (8). We generally prepare 50 or 100 electrode sleeves at a time, perhaps once a year.

Protocol 8. Construction and testing of TPMP-sensitive electrodes (2, 8)

Equipment and reagents

- Five glass Petri dishes
- Polyvinylchloride tubing
- Pt wire soldered to screened cable
- Reference electrode
- pH meter
- Strip chart recorder
- Microsyringes
- Thermostatted and magnetically stirred incubation chamber

- 6 ml tetraphenylboron (10 mM in tetrahydrofuran)
- Polyvinylchloride (1 g added to 20 ml of rapidly-stirred tetrahydrofuran)
- 3 ml dioctylphthalate
- TPMP (10 mM)
- TPMP (1 mM)
- Incubation medium: 120 mM KCl, 5 mM Hepes, 1 mM EGTA pH 7.2

Method

1. Stir the tetraphenylboron and polyvinylchloride together vigorously in a glass vessel and add the dioctylphthalate (plasticizer).

2. Pour the solution into five glass Petri dishes and leave for 24–48 h on a level surface to allow the solvent to evaporate and the membranes to form. At the end of this time the membrane should be colourless and fairly robust.[a]

3. Prepare a number of lengths of polyvinylchloride tubing to form the electrode sleeves.[b]

4. Put a drop of tetrahydrofuran on to the membrane and place a sleeve squarely on the drop. Support the sleeve while the tetrahydrofuran evaporates. Repeat with other sleeves as required.

5. Leave the sleeves for 24–48 h to cure.

6. Cut around the stuck membranes with a sharp blade and carefully remove the sleeve with the attached membrane patch from the Petri dish. Trim the patch of membrane with sharp scissors.

7. Fill the electrode sleeve with 10 mM TPMP (or other cation solution as required) using a syringe. Make sure there are no trapped air bubbles that might prevent proper electrical contact. Leave the sleeve for at least 48 h immersed in 10 mM TPMP.[c]

8. To use an electrode sleeve, insert a length of platinum wire (1–2 cm long) soldered to the core of a screened electrode lead or low noise coaxial cable.[d]

9. Connect the lead, together with a suitable reference electrode, to a suitable voltmeter (e.g. a good quality pH meter) and recorder. The reference of a standard combination pH electrode works well.

10. Check electrode sleeves for sensitive, stable response to TPMP as follows. Immerse the TPMP and reference electrodes in a well-stirred medium containing more than 10 mM salt (e.g. 120 mM KCl, 5 mM Hepes, 1 mM EGTA). Add 1 µl of 1 mM TPMP (1 µM final concentration) and check that there is a response. Add successive 1 µl additions, which should give a smaller response each time. The signal with 5 µM TPMP should be stable and essentially noise-free.[e] Store the TPMP electrode in medium, containing a few µM TPMP if desired, when not in use.

---

[a] The exact thickness of the membrane does not seem to be important for response time, sensitivity, or stability.

[b] We use 4 cm lengths of green earth sleeving, 4 mm outside diameter, from our electrical workshops, but many other tubing types and diameters should be suitable. It is important that the ends should be cut very cleanly—a swift cut with a sharp razor-blade with the tubing under tension works well.

[c] Shorter exposure to TPMP yields electrodes that are less sensitive and are prone to drift. Large numbers of electrode sleeves can be stored for years like this.

[d] The sleeve fits conveniently on to the lead, and a disposable capillary pipette tip inserted into the gap prevents pressure changes that might burst the membrane during assembly or use.

[e] In practice it may take several incubations and TPMP challenges before this stability is achieved. If it is not, try another sleeve. A twofold change in TPMP concentration should change the signal by about 18 mV. In our experience most of the electrodes sleeves in a batch work well, are linear with log [TPMP] down to 1.0 or 0.5 µM TPMP, and improve in drift, stability, and sensitivity during use. Each one will last for months if treated carefully.

3.5.1 Trouble-shooting the construction of TPMP electrodes

(a) The quickest solution to a troublesome electrode sleeve is to discard it and use a fresh sleeve.

(b) Poor electrical screening can produce a very unstable signal. Check the point where the platinum wire is soldered to the screened cable, as this often acts as an aerial. Use well-screened leads or a better quality meter. Earth each of the pieces of apparatus.

(c) Electrical 'spikes' caused by switching of thermostats and other equipment can be reduced by earthing and by running other equipment from a different circuit of the main electricity supply.

(d) Noise in the signal is often caused by the stirring. Check the noise with the stirrer off. If it decreases, adjust the stirring speed and stirrer geometry until you get a quiet trace. The flow of medium past the reference electrode can be crucial—try positioning the reference at different angles to the flow.

(e) Noise in the signal is often caused by the reference electrode, which can become clogged with protein (clean or replace it). Most reference electrodes require 5 or 10 mM salt to give a stable signal.

(f) Freshly made electrodes or those that have been exposed to a recent large change in [TPMP] can show extensive persistent signal changes (drift). In routine use these settle down within a minute or two, but when you first set up an electrode they can sometimes last for days. An electrode that gives a sensitive and stable response to TPMP superimposed on a large drift will usually improve dramatically after storage for several days in medium containing 5 μM TPMP. A few electrode sleeves kept in this way allows trouble-free changeover to a fresh sleeve during a series of experiments.

(g) Slower drifts in the signal that change direction with time are sometimes caused by temperature fluctuations. Check the thermostatting and the fluid flow through the water-jacket.

3.5.2 Use of TPMP electrodes to measure mitochondrial membrane potential

We routinely measure $\Delta\psi$ and respiration rate simultaneously, by inserting the TPMP and reference electrodes through the drilled lid of an oxygen electrode chamber (Rank Bros.). *Protocol 9* describes the use of the TPMP electrode, ignoring the oxygen electrode.

3.5.3 Calculation of mitochondrial membrane potential using a TPMP electrode

It is best to calibrate every trace that is carried out. However, in a long series of traces the calibration is often very similar for each trace, and it can be done less frequently. The calibration tends to change near the start of a series of experiments as the electrode becomes conditioned to TPMP. It also changes quite markedly near the end of the life of an electrode sleeve. *Protocol 10* describes how to calculate the value of the mitochondrial membrane potential.

Protocol 9. Calibration and use of TPMP-sensitive electrodes (2, 4, 5)

Equipment and reagents

- TPMP and reference electrodes
- pH meter
- Strip chart recorder
- Microsyringes
- Thermostatted and magnetically stirred 3 ml incubation chamber
- Mitochondria (50 mg protein/ml)
- Incubation medium: 120 mM KCl, 1 mM

EGTA, 5 mM Hepes pH 7.2, maintained at 37°C
- TPMP (1 mM)
- Rotenone (5 mM in EtOH)
- Nigericin (240 μg/ml EtOH)
- Succinate (1 M)
- FCCP (300 μM in EtOH)

Method

1. Incubate 120 μl of mitochondria in 3 ml of medium (1 mg protein/ml). Add 3 μl (5 μM) rotenone to prevent respiration on endogenous NAD-linked substrates. Add 1 μl nigericin (80 ng/ml) if you want to bring $-z\Delta pH$ close to zero. Insert the TPMP and reference electrodes and wait for a few minutes until the trace is steady.

2. Add 3 μl TPMP (final TPMP concentration of 1 μM). When the trace reaches a new steady value (5–30 sec) add a second 3 μl aliquot of TPMP. Repeat these additions until the total TPMP concentration is 5 μM.

3. Add 30 μl (10 mM) succinate, and allow the mitochondria to accumulate the TPMP for a minute or two until an equilibrium distribution is achieved and the extramitochondrial TPMP concentration is stable.

4. Subsequent additions of inhibitors, ionophores, or other compounds can now be made, with the new steady value of $\Delta\psi$ being achieved within a minute or two of each addition.

5. At the end of the run, add uncoupler (e.g. 1 μl (0.1 μM) FCCP) to dissipate $\Delta\psi$. All of the TPMP should be released by the mitochondria, bringing the external concentration back to 5 μM within a minute.[a]

---

[a] Any drift in the baseline electrode signal can be assessed by comparing the new signal with the original one at the end of step 2. If the drift is mild it is reasonable to construct a baseline assuming constant drift during the experiment. If the drift is severe then the cause of drift must be put right before you attempt to quantify the results using *Protocol 10*.

Protocol 10. Calculation of $\Delta\psi$ from TPMP electrode data

1. Measure the deflection caused by each 1 µM TPMP addition in chart units from the baseline.

2. Plot this deflection against log (final TPMP concentration).

3. To measure the external TPMP concentration for any given electrode signal, measure the deflection from the baseline (in chart units) at the desired steady state. Read off the external TPMP concentration from the calibration graph.

4. If you measure the TPMP binding correction for your conditions using method A (*Protocol 6*), you will need to measure matrix volume using *Protocol 1*. You can then calculate µM [TPMP] in the mitochondrial matrix as {[TPMP] added − external [TPMP]}/{0.001 × matrix volume (in µl/mg protein) × mg protein/ml}.[a]

5. Calculate $\Delta\psi$ as 61.5 log (matrix [TPMP] × TPMP binding correction A/external [TPMP]) (at 37°C).[b]

6. Alternatively, measure TPMP binding correction B for your conditions as described in *Protocol 7*. In this case you need not measure mitochondrial volume for every condition, and can calculate $\Delta\psi$ as 61.5 log ({[TPMP] added − external [TPMP]} × TPMP binding correction B/{0.001 × mg protein/ml × external [TPMP]}) (at 37°C).[c]

7. Average the values from at least three experiments for accurate results.

[a] For example, using mitochondria at 0.9 mg protein/ml, 5 µM added [TPMP], and a measured mitochondrial matrix volume of 0.6 µl/mg protein, then an external [TPMP] of 1.5 µM would correspond to a matrix [TPMP] of {5 − 1.5}/{0.0006 × 0.9} = 6481 µM.
[b] For example, $\Delta\psi$ = 61.5 log (6481 × 0.32/1.5) = 193 mV.
[c] For example, $\Delta\psi$ = 61.5 log ({5 − 1.5} × 0.55/{0.001 × 0.9 × 1.5}) = 194 mV.

3.5.4 Trouble-shooting TPMP electrodes used to measure mitochondrial potential

(a) Check that the electrode sleeve you are using gives a sensitive and stable response to TPMP (see *Protocol 8*). If not, replace it. A sleeve near the end of its life often produces drift, intermittent noise, and decreased sensitivity to TPMP.

(b) The most likely cause of low values for $\Delta\psi$ is that the mitochondria are not properly coupled. Check that they are well coupled under the exact conditions you are using, by observing an increase in respiration rate when uncoupler is added. Too high a concentration of TPMP (more than about 10 µM initial concentration) will uncouple or inhibit respiration. Nigericin can uncouple if there is contaminating valinomycin or a rapid endogenous potassium uniport activity. If the vessel becomes anaerobic then $\Delta\psi$ will collapse.

(c) If the calculated value of $\Delta\psi$ is too high then check the calculations and the measurements of mitochondrial volume and of TPMP binding.

(d) Drift of the electrode trace in the presence but not the absence of mitochondria is sometimes a problem. It is usually caused by the reference electrode (clean it or replace it).

(e) Contamination of apparatus and pipettes with ionophores that slowly leach into the vessel can cause drift in the value of $\Delta\psi$.

4. Measurement of mitochondrial $\Delta\psi$ in intact cells

4.1 Merits of different methods

Although it is possible to measure mitochondrial membrane potential ($\Delta\psi_m$) in intact cells using a TPMP electrode, generally it is better to use radioactive probes for quantitative work. This is because the equilibration time of TPMP in cells is between 10 minutes and 30 minutes so that drifts in the electrode trace can dominate the signals obtained and lead to large inaccuracies. TPP equilibrates a little faster, but this advantage is offset by the greater binding of TPP, which leads to more significant binding corrections and therefore less quantitatively reliable results. Fluorescent probes have also been used with some success, particularly with small amounts of material and with single cells, but once again quantitation of the signal is difficult. Methods to measure mitochondrial ΔpH *in situ* are cumbersome (9, 12) and will not be discussed here.

4.2 Other information needed

Because probes such as TPMP and TPP are accumulated across the plasma membrane as well as across the mitochondrial membrane, it is necessary to know the value of the plasma membrane potential ($\Delta\psi_p$) and the proportion of the cell occupied by mitochondria before $\Delta\psi_m$ can be calculated from cellular TPMP uptake. The main difficulty in measuring $\Delta\psi_m$ is to know the value of $\Delta\psi_p$ sufficiently accurately.

In synaptosomes it is possible to estimate $\Delta\psi_p$ quite accurately using the transmembrane distribution of potassium (measured with ^{86}Rb). In other cells such as lymphocytes and hepatocytes, potassium is not at equilibrium across the plasma membrane, and this ion can not be used. Valinomycin can not be added to whole cells to allow equilibration of ^{86}Rb because the valinomycin will cause both $\Delta\psi_p$ and $\Delta\psi_m$ to be clamped by the relevant potassium gradients and so will set $\Delta\psi_p$ to a high value and will dissipate $\Delta\psi_m$.

In lymphocytes it is possible to measure $\Delta\psi_p$ using TPMP in the presence of an uncoupler such as FCCP, which dissipates $\Delta\psi_m$ but fortunately does not affect $\Delta\psi_p$ (10). In hepatocytes ^{36}Cl distribution can be used since Cl$^-$ is close to equilibrium across the plasma membrane in these cells at plasma membrane potentials smaller than -40 mV (11). For other cell types it would be necessary to validate some way of measuring $\Delta\psi_p$ before proceeding.

To measure $\Delta\psi_m$ you need to measure the cell volume (this can be done using ^{3}H$_2$O and an impermeant probe such as inulin, *Protocol 11*), the plasma membrane potential (*Protocol 12*), the accumulation of TPMP into the cells (*Protocol 13*), the fraction of the cell occupied by mitochondria, and the binding of TPMP to components in the medium, the cytoplasm, and the mitochondrial matrix (6, 12).

The fraction of the cells occupied by mitochondria can be assessed using electron microscopy. Binding of TPMP to the matrix can be assessed in isolated mitochondria using *Protocol 6* and assuming that binding in the matrix in intact cells is similar. Binding of TPMP in the cytoplasm can be assessed using intact cells in the same way, with 1 µg of oligomycin/ml, 1 µM myxothiazol, 20 µM FCCP, and 0.1 µM valinomycin present to dissipate $\Delta\psi_m$ and clamp $\Delta\psi_p$ at the value set by the imposed potassium gradient. TPMP binding in the cytoplasm can be high, but since little of the TPMP in the system is in the cytoplasmic compartment, this binding can often be ignored at high values of $\Delta\psi_m$. Binding of TPMP in the external medium can be avoided by omitting proteins such as serum albumin that bind the probe, or can be quantified by equilibrium dialysis (12).

4.3 Protocols for measurement of $\Delta\psi_m$

The following protocols describe the steps needed to measure $\Delta\psi_m$ in hepatocytes (9, 12, 13). It is wise to add the carrier inulin and TPMP to all experiments to maintain comparability.

Protocol 11. Measurement of hepatocyte volume (12, 13)

Equipment and reagents

- Shaking water-bath
- Automatic pipettes
- Freshly prepared hepatocytes (at about 50 mg dry weight/ml; they can be kept on ice for up to 4 h before use)
- ^{3}H$_2$O (250 µCi/ml)
- [^{14}C]methoxyinulin (10 µCi/ml)
- Carrier inulin (10 mg/ml) warmed to 50°C (no higher!) to dissolve it before use

Method

1. Pre-incubate hepatocytes at 5–10 mg dry weight per ml in a suitable medium[a] in a shaking water-bath (100 cycles/min) at 37°C for 10 min.[b]

2. Add 10 µl (2.5 µCi) of ^{3}H$_2$O and 10 µl (0.1 µCi) of [^{14}C]methoxyinulin per ml, together with 10 µl (0.1 mg) carrier inulin/ml.

---

3. Incubate for 20 min.[c] If uptake or degradation of the inulin is suspected, add the [^{14}C]methoxyinulin 2 min before the end of this incubation.

4. Take three 0.7 ml aliquots. Centrifuge and process them as described for mitochondria in *Protocol 2*.

5. Calculate cell volume in μl/mg dry weight as (^{3}H$_2$O space − [^{14}C]methoxyinulin space)/mg dry weight of cells in 0.7 ml of incubation.

6. Calculate the mean of the values from the three replicates, then average these means from at least three incubations for accurate results. Hepatocyte volumes should be around 1.5–1.7 μl/mg dry weight.

[a] 106 mM NaCl, 5 mM KCl, 25 mM NaHCO$_3$, 0.41 mM MgSO$_4$, 10 mM Na$_2$HPO$_4$, 2.5 mM CaCl$_2$, 10 mM glucose, 10 mM lactate, 1 mM pyruvate, 2.25% (mass/vol.) defatted bovine serum albumin (14), saturated with 95% air/5% CO$_2$ to bring the pH to 7.4.
[b] This pre-incubation is to allow ion gradients to reach their steady state values.
[c] 2 min is sufficient for equilibration of these probes, but the cell volume should be measured after the same incubation time as the other probes.

Protocol 12. Measurement of plasma membrane potential in hepatocytes (11)

The method relies on the *exclusion* of ^{36}Cl$^-$ from the cells. This is necessarily more prone to errors than accumulation methods, so for greater precision the cells are spun through an oil layer to decrease the contamination of the pellet with supernatant.

Equipment and reagents

- Shaking water-bath
- Automatic pipettes
- Bench microcentrifuge
- Scintillation counter
- Vortex mixer
- Hepatocytes
- ^{3}H$_2$O (250 μCi/ml)
- ^{36}Cl$^-$ (10 μCi/ml)
- Silicone oil (42% (v/v) dinonylphthalate and 58% silicone fluid D.C. 550)
- Triton X-100 (2% (v/v) in 250 mM sucrose)
- Scintillation fluid

Method

1. Pre-incubate hepatocytes for 10 min in a suitable medium (*Protocol 11*).

2. Add 10 μl (0.1 μCi) of ^{36}Cl$^-$ and 10 μl (2.5 μCi) of ^{3}H$_2$O per ml.

3. Incubate for 20 min to allow equilibration of ^{36}Cl$^-$.

4. Take three 0.7 ml aliquots. Add them to 1.5 ml microcentrifuge tubes containing 350 μl of silicone oil layered over 100 μl of Triton X-100/sucrose. Centrifuge for 2 min at full speed (about 12 000 *g*) in a microcentrifuge.

Protocol 12. *Continued*

5. Take 200 μl of each supernatant into a scintillation vial and add 3.5 ml of scintillant.

6. Aspirate the residual supernatant and most of the oil layer, then wipe the tube walls with a twisted paper tissue.

7. Resuspend the pellet by vigorous vortex mixing.

8. Cut the base of the tube into a scintillation vial and add 3.5 ml of scintillant.

9. Vortex the vials and count for radioactivity.

10. Calculate pellet spaces as described in *Protocol 2*.

11. Calculate the chloride accumulation ratio (i.e. $[Cl^-]_{in}/[Cl^-]_{out}$) as $[(^3H_2O$ space $- [^{14}C]$methoxyinulin space$) - (^3H_2O$ space $- {}^{36}Cl$ space$)/(^3H_2O$ space $- [^{14}C]$methoxyinulin space$)]$. Use values for $(^3H_2O$ space $- [^{14}C]$methoxyinulin space$)$ determined in parallel using *Protocol 11*.

12. If required, calculate plasma membrane potential as 61.5 log (chloride accumulation ratio). For hepatocytes the value should be about -28 to -35 mV.

13. Average the values from the three replicates, and average these mean values from at least six preparations for accurate results.

Protocol 13. Measurement of mitochondrial membrane potential in hepatocytes using TPMP (12, 13)

Equipment and reagents

- Shaking water-bath
- Automatic pipettes
- Scintillation counter
- Vortex mixer

- Hepatocytes
- [^{3}H]TPMP (100 μCi/ml)
- TPMP (0.1 mM)

Method

Measure cell volume (*Protocol 11*) and plasma membrane potential (*Protocol 12*) in parallel.

1. Pre-incubate hepatocytes in a suitable medium for 10 min (*Protocol 11*).

2. Add 10 μl (1 μCi) of [^{3}H]TPMP per ml, together with 10 μl (1 μM) carrier TPMP.

3. Incubate for 20 min to allow equilibration of the probe.

4. Take three 0.7 ml aliquots. Centrifuge and process them as described for mitochondria in *Protocol 2*.

5. Calculate TPMP accumulation ratio (i.e. $[TPMP]_{in}/[TPMP]_{out}$) as ($[^3H]$TPMP space $-$ $[^{14}C]$methoxyinulin space)/(^3H_2O space $-$ $[^{14}C]$methoxyinulin space).[a]

6. Calculate $\Delta\psi_m$ using the following equation:

$$\Delta\psi_m = 61.5 \log \left| \frac{V_c \cdot a_m}{V_m \cdot a_c} \left[\frac{[Cl^-]_{in} \cdot [TPMP]_{in} \cdot a_c (V_c + V_m)}{(Cl^-]_{out} \cdot [TPMP]_{out} \cdot a_e \cdot V_c} - 1 \right] \right|$$

where V_c and V_m are the proportions of the cell volume occupied by cytoplasm and by mitochondria (0.8 and 0.2 for hepatocytes), a_m, a_c, and a_e are the apparent activities (binding corrections) of TPMP in the mitochondrial matrix (0.4), cytoplasm (0.2), and external medium (0.7), and subscripts 'in' and 'out' refer to the probe concentrations in the cells and in the medium (see refs 12 and 13). $\Delta\psi_m$ is usually between 140 and 160 mV in resting cells.

7. Average the values from the three replicates, and average these mean values from at least three preparations for accurate results.

---

[a] At high potentials the $[^{14}C]$methoxyinulin space is negligible compared to the $[^3H]$TPMP space and need not be subtracted from it.

4.4 Trouble-shooting measurements of $\Delta\psi_m$ in cells

(a) Poor cell viability (less than 90% excluding trypan blue), lack of oxygen, inadequate control of incubation pH, contamination with ionophores and inhibitors, or overly long incubations can lead to increases in $\Delta\psi_p$ and decreases in $\Delta\psi_m$.

(b) If the calculated cell volumes are negative or too small, check that the inulin is fully dissolved and not contaminated with bacteria–precipitation of inulin will overestimate the inulin space and underestimate the cell volume. Prior removal of precipitable material by centrifugation is sometimes useful. Ensure that ^3H_2O is not being lost during processing of samples; this will cause the ^3H_2O space and the cell volume to be underestimated.

(c) Because the $[^3H]$TPMP is greatly accumulated, even a small contamination with ^3H_2O (which is not accumulated) will have a large effect on the calculated ratio $[TPMP]_{in}/[TPMP]_{out}$.

(d) Errors in $\Delta\psi_p$ result in similar mV errors in $\Delta\psi_m$, but twofold errors in V_m, V_c, and TPMP binding tend to give only 18 mV difference in $\Delta\psi_m$.

Acknowledgements

I thank Dr R. K. Porter for helpful comments on the manuscript.

References

1. Nicholls, D. G. and Ferguson, S. J. (1992). *Bioenergetics 2*. Academic Press, London.
2. Brown, G. C. and Brand, M. D. (1985). *Biochem. J.*, **225**, 399.
3. Brand, M. D. (1985). *Biochem. J.*, **225**, 413.
4. Murphy, M. P. and Brand, M. D. (1987). *Biochem. J.*, **243**, 499.
5. Hafner, R. P., Nobes, C. D., McGown, A. D., and Brand, M. D. (1988). *Eur. J. Biochem.*, **178**, 511.
6. Brand, M. D. and Felber, S. M. (1984). *Biochem. J.*, **217**, 453.
7. Brown, G. C. and Brand, M. D. (1988). *Biochem. J.*, **252**, 473.
8. Kamo, N., Muratsugu, M., Hongoh, R., and Kobatake, Y. (1979). *J. Membr. Biol.*, **49**, 105.
9. Hoek, J. B., Nicholls, D. G., and Williamson, J. R. (1980). *J. Biol. Chem.*, **255**, 1458.
10. Felber, S. M. and Brand, M. D. (1982). *Biochem. J.*, **204**, 577.
11. Nobes, C. D. and Brand, M. D. (1989). *Biochim. Biophys. Acta*, **987**, 115.
12. Nobes, C. D., Brown, G. C., Olive, P. N., and Brand, M. D. (1990). *J. Biol. Chem.*, **265**, 12903.
13. Harper, M.-E. and Brand, M. D. (1993). *J. Biol. Chem.*, **268**, 14850.
14. Chen, R. F. (1967). *J. Biol. Chem.*, **242**, 173.

4

Reconstitution of bioenergetic proteins and the uses of proteoliposomes

JOHN M. WRIGGLESWORTH

1. Introduction

A variety of membrane proteins are involved in bioenergetic processes such as ATP synthesis, ATP hydrolysis coupled to ion and metabolite translocation, and redox reactions coupled to proton and sodium ion pumping. A study of their mechanism of action is often aided by purification and reconstitution into artificial bilayer systems, the most widely used of which is the liposome (1, 2). The advantage of reconstituted systems is that vectorial reactions of the protein can be recreated under defined conditions. The disadvantages are largely methodological. Problems can arise in the choice of reconstitution method, the lipid composition of the bilayer for optimum activity, and the yield and homogeneity of the final preparation. In addition, the properties of the vesicles themselves can impose constraints on the function of the protein. For example, the small vesicle size produced by some methods of preparation means that the electrogenic movement of just a few charged molecules can generate a significant membrane potential.

In the present chapter, I describe in detail the most common ways of making and characterizing proteoliposomes (here defined as liposomes with a transmembrane incorporation of the protein or proteins under study). The methodology for proteoliposome production is very restricted compared to that used for making liposomes of lipid alone (see ref. 3 for liposome examples), mainly because of the sensitivity of protein to denaturing conditions such as exposure to organic solvent, mechanical stress, or extreme variations of pH. Unfortunately, even when these conditions are avoided, it is still not possible to predict which method will be the most suitable for any particular protein. In many instances it is a case of 'try it and see'. However, since the methodology is relatively simple and quick, this strategy is not too much of a disadvantage.

2. Making proteoliposomes

2.1 The protein

Membrane proteins, by their very nature, require detergents for their solubilization and purification. The final purified fraction often needs relatively high (> 1%) concentrations of detergent to stabilize the protein. This can be a problem for reconstitution where the detergent is effectively replaced by lipid. Residual detergent can alter the permeability properties of the vesicles or even prevent their formation. For this reason, the final purified sample of protein should be prepared in as concentrated a form as possible, in a detergent concentration as low as is compatible with stability. This minimizes the volume of protein (and hence detergent) added to the reconstitution mixture. A further help is to suspend the final protein fraction in a detergent with a high critical micelle concentration (see *Table 1*). Residual detergent can then be removed during or after reconstitution by dialysis. Because of their low critical micelle concentrations, polyethylene glycol detergents (for example Triton X-100) should be avoided. If this is inescapable, then gel permeation or specific adsorption methods will have to be used for removal (4).

Experience has shown that for best reconstitution, the membrane protein should be of high purity. Contaminating protein can interfere with the incorporation process, presumably by forming non-specific aggregates and preventing correct transmembrane insertion.

2.2 Choice of lipid

Surprisingly, at first sight, the lipid composition for reconstitution of many membrane proteins is not very stringent. The reason for this is that specific lipids are usually tightly bound to the protein and often remain bound until

Table 1. Critical micelle concentrations (CMCs) of various detergents used in protein purification and reconstitution studies

Detergent	CMC	
	mM	% (w/v)
Octyl glucoside	23	0.07
Sodium cholate	14	0.06
Sodium deoxycholate	4	0.02
Decyl maltoside	2.2	0.01
Dodecyl maltoside	0.2	0.0016
Polyoxyethylene types		
Triton X-100	0.2	0.0015
Tween 20	0.03	0.0004
Brij 98	0.025	0.0004

the final stages of purification (5). An example is the cardiolipin associated with cytochrome oxidase (6). Many membrane proteins become unstable if fully delipidated since few detergents can match the specific lipid requirements of a particular protein. The successful reconstitution of the acetylcholine receptor requires specific phospholipids to be added during the purification of the protein to maintain its integrity (7). It is recommended therefore that, unless the specific lipid–protein interaction itself is under study, the reconstituted protein should not be fully delipidated. Then, in most cases, all that is required is to provide a suitable lipid for bilayer formation and insertion of protein.

Most proteins require the phospholipids to have some unsaturation for successful reconstitution. If specific acyl chains are used, reconstitution should be done above the transition temperature of the lipid. A crude extract of soybean lipid (sold as L-α-phosphatidylcholine, Sigma type IV-S) which contains varying amounts of other lipids can often be used for initial experiments. A detailed analysis of the composition of soybean lipid is given in ref. 1. Phospholipids from soya have higher content of linoleic acyl chains, and hence higher degree of unsaturation, than those from egg. The advantages of using phospholipids from soybean for initial experiments are that it supports a high activity in a variety of membrane enzymes and is inexpensive. If specificity is required for optimum protein activity, then more defined phospholipid mixtures can be supplemented with lipid extracted from the original membranes containing the protein. In experiments that may involve affinity adsorption of the reconstituted protein in order to purify a specific proteoliposome fraction (see section 3.2.1), a more careful choice of lipid head group may be necessary. For example, Madden and Cullis (8) have separated proteoliposomes of cytochrome oxidase with an outwardly oriented cytochrome *c* binding site on an affinity column using dioleoylphosphatidyl choline as the lipid for reconstitution. When a small amount of phosphatidyl serine was included in the bilayer, adsorption became much poorer because of the interfering negative charge on the lipid.

The presence of non-esterified fatty acids should be avoided wherever possible. These can act as protonophores and even catalyse the movement of monovalent and divalent cations across the lipid bilayer (9). In crude phospholipid mixtures, these can be removed by acetone washing as described by Darley-Usmar *et al.* (10).

2.3 Lipid/protein ratio

One of the puzzling findings of reconstitution studies has been that vectorial activities are best retained at relatively high lipid to protein ratios, usually 50:1 by weight (molar ratios typically around 5000:1), which contrasts to the much lower ratios found in biological membranes. The molar ratios of several thousand suggest that reconstitution is best when one molecule of protein is incorporated into one vesicle. A single vesicle of diameter 50 nm would contain around 5000 molecules of phospholipid. In support of this, Madden *et al.* (11) report that the optimum respiratory control for cytochrome oxidase proteoliposomes occurred when one oxidase complex was incorporated per vesicle. However, Tihva *et al.* (12) found that larger vesicles can be made with appreciable amounts of incorporated cytochrome oxidase. As long as high lipid/protein ratios are maintained, the proteoliposomes still show high respiratory control values. The number of incorporated protein molecules per vesicle matters less than the overall lipid/protein ratio. The reasons for this are not known but differential scanning calorimetry studies have shown that the co-operativity of phospholipid melting is profoundly disrupted by the incorporation of small amounts of purified membrane glycoprotein into bilayers (13). This seems to have a dramatic effect on the phospholipid gel to liquid crystalline transition. For membrane glycoprotein, the measure of co-operativity was found to be halved at a lipid/protein molar ratio of 3800:1 and protein clusters were seen by freeze–fracture electron microscopy at lower ratios (14). It would appear that protein–protein interactions in the natural membrane are under a close control which is lost in the proteoliposome system.

2.4 Methods of formation

2.4.1 Sonication

The use of sonication to prepare proteoliposomes is probably the simplest and quickest of reconstitution procedures. However, an empirical approach has to be taken because many parameters can affect the sonication process (probe size, power output, sample volume, etc.). Local heating is necessarily high in the sonication process and the energy has to be dissipated efficiently if bulk heating and protein denaturation is to be avoided. A bath sonicator would be a good choice from this viewpoint but energy transfer to the suspension is often much lower than with probe sonicators, resulting in poor reconstitution. I have found that a probe sonicator, operating in a pulse mode where energy is transferred in short pulses (typically one-third of a second on and two-thirds of a second off) is an efficient sonication method. The sample is best placed in a glass container for efficient heat transfer to a surrounding ice–water mixture but care must be taken to choose unflawed glass vessels and to keep the probe away from any contact with the glass. Old fine-chemical bottles can be conveniently used for 5–10 ml samples. If plastic containers have to be used, then the cooling time between short bursts of sonication should be extended. Another advantage of glass containers is that the position of the probe in the sample can be seen. This should be far enough into the liquid to avoid emulsification with air dragged down from the surface meniscus during sonication. When this happens, the power transfer efficiency drops dramatically (a change in the 'hard' sound of sonication will become immediately obvious, even through protective ear shields which should always

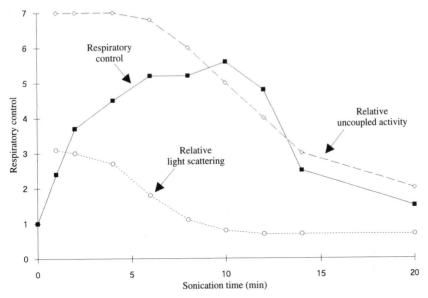

Figure 1. Monitoring proteoliposome formation. Ideally, three parameters need to be monitored during a reconstitution procedure; vesicle formation, protein incorporation, and enzyme activity. In this example, the sonication procedure is being used to reconstitute cytochrome oxidase proteoliposomes. Vesicle formation is monitored by absorbance measurements at 450 nm which give a measure of the relative size of the vesicles from their light scattering properties. Incorporation of cytochrome oxidase into the membrane is monitored by the ability of the proteoliposomes to exhibit respiratory control. (The vectorial generation of a proton motive force inhibits cytochrome oxidase activity unless relieved by the presence of a protonophore (34). Respiratory control is the ratio of activity in the presence of a protonophore to that in its absence.) Finally the uncoupled activity of the enzyme (activity in the presence of the protonophore) is monitored to assess the damaging effect of sonication, in this case occurring progressively after 6 min.

be worn). The fine mist of spray from the liquid during sonication can be contained by a loose wrap of Parafilm around the neck of the container.

The process of reconstitution can be checked in a separate experiment by removing samples of the mixture every few minutes and measuring absorbance (at a wavelength away from any absorption band of the protein) to monitor liposome size, together with a parameter characteristic of the activity of the reconstituted protein. An example for the formation of cytochrome oxidase proteoliposomes is described in *Protocol 1* and shown in *Figure 1*. The absorbance at 470 nm falls with time to a constant level with the formation of small uni- and oligolamellar vesicles. Over the same time period, the respiratory control of the proteoliposomes rises but eventually denaturation occurs and the activity of the enzyme falls. In this example, an optimum time of sonication would be around eight to ten minutes.

Protocol 1. Preparation of cytochrome oxidase proteoliposomes by sonication

Equipment and reagents

- Pulse sonicator with 30% duty cycle
- Sephadex G-25 column (for preparation of vesicles containing entrapped probe molecules)
- Solid asolectin (phosphatidylcholine type IV-S)—alternatively a mixture of phosphatidylcholine (egg) and phosphatidyl ethanolamine (egg) in weight proportions 1:1,

dissolved in chloroform–methanol can be used
- Potassium phosphate buffer, 100 mM pH 7 (other buffer mixtures can be used)
- Purified cytochrome oxidase (about 200–300 μM with haem a : protein ratio greater than 8 nmol/mg suspended in 100 mM K⁺-phosphate, 0.5% Tween 80)

Method

1. Weigh 0.25 g asolectin into the container to be used for sonication. Dissolve in 0.5 ml chloroform and dry down under nitrogen. Alternatively dry down the mixture of purified phospholipids under nitrogen.

2. Add 5 ml buffer solution and vortex thoroughly for several minutes. [a]

3. Add 5 mg cytochrome oxidase (around 150 μl) and re-vortex for 30 sec.

4. Sonicate for 5–10 min in a pulse sonicator with 30% duty cycle. Keep the mixture cool using a surrounding ice–water mixture.

[a] For vesicles containing entrapped probe molecules, such as the pH-sensitive dye pyranine, add the probe at step 2 to the final concentration required (usually mM range). Separate non-entrapped probe from the vesicles after sonication by gel chromatography using a Sephadex G-25 column.

2.4.2 Detergent dialysis

In this method, the purified protein is dispersed in a suitable mixture of detergent and phospholipid which is then slowly dialysed to remove the detergent (see *Protocol 2*). Because dialysis requires relatively large volumes, this method can be expensive if probe molecules are to be entrapped inside the vesicles. Equal concentrations of probe have to be present in the dialysis medium. Efficient dialysis is essential (15). Formation of the vesicles requires that the mixture is continuously agitated. This can be done by including the dialysis bag and dialysate inside a roller bottle and leaving on a roller apparatus during dialysis.

Some enzymes are particularly sensitive to cholate and other detergents should be used before abandoning the technique altogether. Octyl glucoside in particular has been used successfully with a wide variety of proteins. A variation on the detergent dialysis procedure given in *Protocol 2* is to simply dilute the mixture after step 2. Because relatively high detergent concentrations are needed to solubilize the initial mixture, a dilution of around 100–fold is necessary (16).

Protocol 2. Preparation of cytochrome oxidase vesicles by cholate dialysis

Equipment and reagents

- Phospholipid, cytochrome oxidase, and buffers (as in *Protocol 1*)
- Sodium cholate (solution should be colourless, otherwise re-crystallization is recommended[a])

- Dialysis tubing (1 cm diameter) pre-boiled in 10 mM EDTA for 10 min, cooled, and washed in distilled water
- 500 ml round culture bottle and Universal roller apparatus

Method

1. Disperse 0.25 g of the phospholipid in 5 ml buffer containing 0.1 g sodium cholate. Sonicate for 30 sec to ensure full dispersal.

2. Add cytochrome oxidase to a final concentration of around 4 μM and mix.[b]

3. Load into dialysis tubing and place inside a roller bottle together with 500 ml buffer. Dialyse, on a roller apparatus, with changes of buffer at 2, 4, and 15–20 h.

[a] See ref. 9.
[b] If probe molecules are to be entrapped, these should be added at step 2. The same concentration of probe should be present in the buffer throughout dialysis. To reduce cost, 1 ml of sample could be dialysed against 100 vol. of buffer plus probe and the last dialysate could be saved to be used as the first dialysate in the next preparation. Separate non-entrapped probe from the vesicles as described in *Protocol 1*, step 5.

2.4.3 Freeze–thaw

Freeze–thaw cycles provide a conveniently rapid method for the formation of proteoliposomes (17). The technique is usually combined with brief sonication bursts before each freeze cycle (see *Protocol 3*). The method produces larger vesicles than the sonication or dialysis procedures but in a more heterogeneous population with regard to size and numbers of incorporated protein molecules per vesicle. Removal of any detergent used to solubilize the protein can be achieved by a suitable choice of column chromatography.

Protocol 3. Preparation of cytochrome oxidase proteoliposomes by the freeze–thaw technique

1. Prepare 1 ml of a mixture of phospholipid and cytochrome oxidase as in *Protocol 1*, steps 1–3 but using a small plastic container.

2. Sonicate in a pulse mode for a total of 30 sec using a microtip on the sonicator, keeping the mixture cool.

3. Freeze the suspension either in liquid nitrogen or in a dry ice–ethanol bath. Freezing should be as rapid as possible.

Protocol 3. *Continued*

4. Thaw in a water-bath at room temperature (about 10–15 min) and re-sonicate for 10 sec.

5. Repeat the cycle two more times.

6. Pass down a Sepharose 2B column equilibrated with the same buffer as used in the liposome mixture. Discard the void volume fraction (large multilamellar aggregates containing some oxidase but with little respiratory control). Collect the remaining material in one fraction.

At first sight it would seem that the freeze–thaw technique has little to offer in comparison with the sonication or detergent dialysis methods. However, it does avoid the use of prolonged sonication or exposure of the protein to high detergent concentrations. It also provides a simple rapid way of entrapping water soluble molecules inside preformed proteoliposomes (18). These can be added to the mixture before freezing. Non-entrapped material is removed at the gel filtration stage and the use of preformed vesicles reduces the amount of waste material in the void volume.

A further advantage as a technique is that freeze–thaw cycles can be used to fuse preformed proteoliposomes with other vesicles or even membrane preparations. For example, proteoliposomes of cytochrome oxidase prepared by the dialysis method using octyl glucoside have been fused with purified yeast plasma membranes (19). The resulting hybrid vesicles were used to study the influence of protonmotive force on galactose transport by the galactose-specific transporter (see section 5).

2.4.4 Extrusion

An advance in the methodology of liposome formation has been the use of filters with defined pore sizes (20, 21). Repeated extrusion under moderate pressures of multilamellar vesicles through polycarbonate filters with defined pore size results in a relatively homogeneous population of unilamellar vesicles in a short (< 10 min) time. The heterogeneity of the preparation depends on several factors including lipid composition, the pressure used for extrusion, and the number of passes used. A detailed description of the method is given by New (3). Filters with different pore size, ranging from 30–400 nm can be used. These are available from Nuclepore Corp. A standard pressure filtration apparatus, such as that used for concentrating purified protein samples (Amicon Corp.), can be used but custom built extrusion devices have been built for repeated cycling of the liposome material through the filter (Lipex Biomembranes). The technique has not been used to the same extent for reconstitution studies. One restriction is the need for preformed multilamellar vesicles before starting the extrusion process. There are few studies on the extrusion of preformed proteoliposomes. Usually the phospholipid vesicles

formed by extrusion are used for fusion, either with preformed proteoliposomes or with purified protein (22).

2.4.5 Fusion

Fusion of preformed proteoliposomes is a useful technique for preparing
single vesicles incorporating two separate membrane proteins (see section 5).
As mentioned above, the freeze–thaw technique has been used successfully
to fuse cytochrome oxidase proteoliposomes with yeast plasma membrane
preparations. Fusion of purple membrane fragments from *Halobacterium
halobium* with vesicles of phosphatidylcholine has been induced by adding
short chain lecithins such as diheptanoyl phosphatidylcholine at 50 mM followed
by vortexing and brief sonication (23). The molar ratio of phosphatidylcholine
to bacteriorhodopsin was between 320 to 990 and the mixture was maintained
at a temperature above the lipid phase transition of the respective long chain
lecithins. A low pH (pH 5) procedure has been used to induce fusion between
S. cremoris membrane vesicles and proteoliposomes incorporating bacteriorhodopsin (24). Non-specific phospholipid transfer protein has also been used
to promote net transfer of lipid from preformed vesicles of dimyristoyl phosphatidylcholine into purple membrane (25).

3. Characterization of proteoliposomes

3.1 Size and structure

There is as yet no single reliable method for the accurate size distribution and
structure determination of proteoliposome samples. A combination of
methods should be used if full characterization is required.

3.1.1 Microscopy

Structural information on proteoliposomes mainly comes from freeze–
fracture and negative stain electron microscopy. Unfortunately, neither can
give a full picture for most proteoliposome samples. The samples have to be
treated for an image to be made visible under the electron beam and this
process can introduce many artefacts (26). Each sample preparation technique has its limitations. For example, claims for unilamellar structure are
often made on the basis of freeze–fracture appearance whereas negative
staining of the same preparation might easily show oligomeric structures (27).
The reason is that in freeze–fracture, the probability of cross-fracture to
reveal a number of lamellae is low for small vesicles. Both preparation techniques are difficult to apply to minimal size vesicles. Very few attempts have
been made to prepare thin sections mainly because of the loss of phospholipid
during dehydration. Some success is reported by using the water soluble
resin, glycol methacrylate (27).

Despite these problems, samples for negative stain examination are
easily prepared and can be stored almost indefinitely. A small drop of the

proteoliposome sample should be mixed with an equal volume of 2% ammonium molybdate and allowed to dry on a grid coated with polyvinyl formal (formvar) or nitrocellulose (collodion). Alternatively the sample can be sprayed directly on to the grid. Fast blotting should be avoided as this can lead to an unequal size distribution on the grid. The support grids are hydrophobic and spreading can be helped by mixing the sample with serum albumin (1%) if this is known not to affect the sample. Sugar solutions unfortunately give rise to numerous artefacts and the method is not recommended for proteoliposomes prepared in sucrose or after sucrose density gradient centrifugation.

The freeze–fracture technique is especially useful for visualizing intramembrane proteins. On fracture, the fracture plane in a frozen specimen will follow the line of least resistance, which at low temperature will be along the hydrophobic region in the centre of the membrane. Fracture will expose any transmembrane proteins either as particles or pits in the replicated surface. To avoid artefacts of freezing such as ice crystal formation with subsequent concentration increases in water soluble molecules, freezing should be as fast as possible. Cryoprotectants should be avoided. Liquid nitrogen is not particularly fast as a freezing method since a layer of nitrogen gas soon forms around the sample and its low thermal conductivity impedes fast freezing. Liquid nitrogen 'slurry' (at $-210\,°C$) can be used by placing liquid nitrogen under vacuum for a few minutes to cool it to its freezing point. Alternatively, very fast freezing rates can be achieved using liquid Freon-22 or propane, and also by spray-freezing on to a pre-cooled copper block. Shadowing should be as light as possible to maintain fine detail. Tantalum/tungsten shadowing may give higher resolution than the more normal platinum/carbon method. Using the freeze–fracture technique, Tihova *et al.* (12) have analysed cytochrome oxidase proteoliposomes and distinguished incorporated enzyme dimers from higher aggregates. They have also identified monomer structures of subunit III depleted enzyme in the membrane.

3.1.2 Trapped volume

Trapped volume is a parameter easily measured and, together with information from electron microscopy, can be used to fully characterize the size and structure of the proteoliposomes. The basic method is to entrap an impermeant water soluble material at a known concentration within the proteoliposomes when they are formed and to remove any non-entrapped material after formation. A known amount of vesicles can then be solubilized and the concentration of the released material measured. Trapped volume can then be calculated and related to the amount of phospholipid in the sample. Phosphate buffer is a convenient material to use. The buffer can be entrapped at 100 mM for example and the vesicles when formed can be passed down a gel column in 70 mM potassium sulfate (an osmotic match at pH 7). The inorganic phosphate concentration of a known amount of solubilized vesicle preparation can then be measured. A simple ratio calculation, assuming the internal entrapped concentration remained at 100 mM phosphate after vesicle formation, can give the trapped volume in litre per mole phospholipid. Small unilamellar vesicles have a value around 0.4 litre/mole phospholipid. Values for larger liposomes, for example those formed by reverse phase evaporation (without incorporated protein), range between 2–5 litres/mole phospholipid whilst the largest unilamellar liposomes, formed by the ether evaporation technique (again without incorporated protein), have trapped volumes greater than 20 litres/mole phospholipid. Since trapped volume is an average parameter for all the vesicles, the presence of multiple lamellae will affect the calculation of size. Electron microscopy (negative stain) should therefore be used to assess the degree of multiple bilayer forms before size calculations are done from volume measurements.

3.1.3 Light scattering

Particle size distribution analysis by light scattering is a well established technique in the polymer and pharmaceutical industries but has been less well applied to liposome technology. However, with the development of photon correlation spectroscopy and the wider availability of automatic instruments (for example the Malvern Autosizer, Malvern Instruments), a rapid and reliable method of size measurement from a few tens of nanometers up to a micron is now possible. The method is based on measurements of fluctuations of light scattered from suspended particles illuminated by a coherent light beam from a laser. Analysis of the time variable intensities together with a knowledge of the viscosity, refractive index, and temperature of the suspending medium can give a measure of the mean size of the particles (28, 29). Normally, photon correlation spectroscopy measurements are made at low concentrations of sample, typically around 100 μg/ml phospholipid or less, and take only a few minutes. However the sample has to be free of contaminating scattering material and multiple scattering can interfere with the analysis when particle size becomes greater than 1 μm. The number of lamellae per vesicle has to be determined separately by microscopy although some indication of the presence of multilamellar structures can be gained from a comparison of trapped volume measurements with mean vesicle size.

3.2 Homogeneity of the preparation

A preliminary fractionation procedure to produce a more homogeneous sample based on size is often useful. Centrifugation should be used to remove large lipid aggregates and any titanium fragments from the sonicator probe if sonication was used in reconstitution; 15 min at 30 000 *g* is sufficient. For a more quantitative fractionation by centrifugation, higher speeds and longer times are required. Small unilamellar vesicles, such as produced by the detergent dialysis procedure can be purified by centrifugation at 160 000 *g* for

3–4 h (30). Longer times and the use of density gradients have been used to separate proteoliposomes on the basis of size and amount of incorporated protein (31–33), but a less tedious method for size fractionation is the use of gel permeation chromatography. The availability of a wide selection of gel materials, such as the Sephacryl series (Pharmacia Biotech Ltd.), has simplified the preparation of uniformly sized proteoliposomes. Entrapped dye material, such as pyranine or phenol red, can be used to monitor the progress of the vesicles down the column. Alternatively, absorbance measurements of the eluted fractions can be used to monitor light scattering.

3.2.1 Protein orientation

Orientation of incorporated protein in the membrane of reconstituted proteo-liposomes depends strongly on the method of reconstitution. It is not possible to predict for an unknown system which way a membrane protein will orient although packing constraints in minimal sized vesicles will tend to force wedge-shaped proteins to orient with the larger volume to the outer surface. For example, a random (50/50) orientation of cytochrome oxidase occurs when reconstitution by sonication is used, in contrast to the detergent dialysis method where greater than 80% of the oxidase molecules orient so as to expose their cytochrome c reaction sites to the external medium (27).

In any reconstituted system, some form of vectorial assay is required to assess orientation of the protein. Usually this depends on the impermeability of the membrane to a substrate, for example ATP, NADH, or cytochrome c, so that only enzyme molecules able to react with the externally added substrate will be measured. This can then be compared to the activity of the enzyme when the vesicular structure is destroyed, for example by detergent. Unfortunately the intrinsic effects of many detergents on enzyme activity can complicate the calculations. Alternatively, some other vectorial characteristic of the enzyme can be exploited. For example, if externally added ascorbate plus cytochrome c is used as substrate for cytochrome c oxidase incorporated into proteoliposomes, then when the oxygen concentration falls to zero, those molecules with their cytochrome c binding sites exposed to the external medium will become fully reduced and their cytochrome spectrum can be measured. This can be compared to the spectrum of the vesicles when a membrane permeable electron donor such as $N,N,N,'N'$-tetramethyl-p-phenylenediamine (TMPD) is added to the system and all the oxidase molecules become reduced (34). Nicholls *et al.* (35) have used selective inhibitors of varying membrane permeability to assess the orientation of cytochrome c oxidase in proteoliposomes. Impermeant labelled probes, reacting to specific sites in the protein could also be employed. Devising an assay to measure protein orientation for a particular system can be an enjoyable challenge to the researcher.

A more homogeneous preparation of reconstituted vesicles can be isolated from a heterogeneous preparation using affinity chromatography to select vesicles with particular protein orientation in the membrane. A simple example is the use of DEAE–Sephacryl chromatography to adsorb vesicles incorporating cytochrome oxidase with the cytochrome c binding site exposed to the exterior (12, 36). The negatively charged groups around the cytochrome c binding site on the oxidase interact with the positively charged groups on the resin. Higher specificity can be achieved using a cytochrome c affinity column (8).

4. Measurement of ion movements

4.1 The problems of size and capacitance

The electrical properties of a phospholipid vesicle will be determined to a large extent by its size. Assuming the vesicle to behave like a spherical condenser, then its electrical capacitance, C, can be expressed as:

$$C = \frac{Kab}{b - a} \text{ cm}$$

where K is the dielectric constant of the material in the bilayer, a is the radius of the inner sphere, and b the radius of the outer sphere (27). The units have to be carefully watched in calculations. 1 cm is equivalent to 1 electrostatic unit (esu) of capacitance, which in turn is equivalent to 1.1×10^{-12} F. A vesicle with an external diameter of 30 nm with a bilayer thickness of 3.7 nm would have a capacitance of around 1.5×10^{-17} F if K is assumed to have a value of 3 (37). These calcualtions fit well with measured values of membrane capacitance of around 1 μF/cm^2 (38). The significance of these calculations becomes apparent for small vesicles where it can easily be calculated that the electrogenic movement of a few ions can generate an appreciable opposing membrane potential (39).

The small trapped volumes of many proteoliposomes should also be remembered when considering the concentration of entrapped molecules. A unilamellar vesicle of 30 nm diameter will have an internal volume of 5×10^{-21} litres. This means that it only requires three molecules of a substance to have an effective concentration of 1 mM. The statistical fluctuation in concentration of the substance at the molecular level, as the vesicles are formed and the substance is entrapped, will result in a heterogeneous preparation of vesicles with varying internal concentrations of the substance. To reduce this variation, higher concentrations of the entrapped material should be used and/or larger vesicles prepared.

4.2 Measurement of ΔpH

The simplest method for measuring pH gradients across liposomal membranes is to use an entrapped dye which is sensitive to pH. The dye should be freely water soluble to avoid any partition into the membrane, should be charged in both protonated and unprotonated forms to prevent any leakage, and should have a pK_a around the pH of interest. Phenol red was one of the first such dyes to be used (40) but more polar indicators have been developed, one of these being pyranine (8-hydroxy-1,3,6-pyrene trisulfonate, Molecular Probes Inc.). This can be entrapped inside the proteoliposome (see *Protocols 1* and *2*) at a concentration of 5–10 mM and changes in absorbance monitored at 460 nm or fluorescence at 540 nm (excitation at 460 nm). Any dye chosen for ΔpH measurements should be checked to see if incorporation into phospholipid vesicles alters its pK_a. There is a small (approx. 0.2 pH) units change for phenol red ($pK = 7.8$) but no significant change for pyranine ($pK_a = 7.55$). Calibration of ΔpH can be done by the addition of known amounts of H^+ and OH^- added as H_2SO_4 or KOH respectively, in the presence of nigericin to allow equilibration of H^+ and K^+. A typical trace for ΔpH measurement, with detailed experimental conditions, is shown in *Figure 2*. One problem with using entrapped pH indicators to monitor movement of H^+ catalysed by incorporated protein is that only those proteoliposomes with incorporated enzyme will contribute to the pH changes. On the other hand, all the vesicles, whether incorporating enzyme or not, will respond following the addition of acid and alkali for calibration. Heterogeneity in the preparation can therefore lead to an underestimate of the active H^+ changes (see ref. 39 for a discussion of this problem).

An alternative approach to the determination of transmembrane pH gradients is to measure the distribution of weak acids or bases across the vesicle bilayer (41, 42). The principal of the method is that the neutral form of the probe will readily permeate the bilayer and an equilibrium of the protonated and non-protonated forms will be established on both sides of the membrane. Thus:

$$\frac{[\text{probe}^+]_{in}}{[\text{probe}^+]_{out}} = \frac{[H^+]_{in}}{[H^+]_{out}}$$

assuming that only the protonated form can cross the membrane. If the external pH is known, then it only remains to measure the ratios of entrapped probe to free probe to be able to calculate the internal pH. This has been done using a number of techniques including NMR (43), EPR (44), fluorescence (45), and ion-selective electrodes (46), as well as radiolabelling measurements. The amount of trapped probe is usually determined by a speedy removal of the vesicles from the suspending medium. This is usually done by fast centrifugation through an appropriate filter in order to reduce the efflux of trapped probe, which will occur as the equilibrium between internal and external probe is disturbed. To calculate concentrations, the trapped volume of the vesicles also needs to be measured. Since this is an average measure for all vesicles, whether the protein under study is incorporated or not, the calculated ΔpH will be subject to the same problem of heterogeneity as mentioned above. In addition, the method does not permit continuous monitoring of ΔpH as do the spectrophotometric methods.

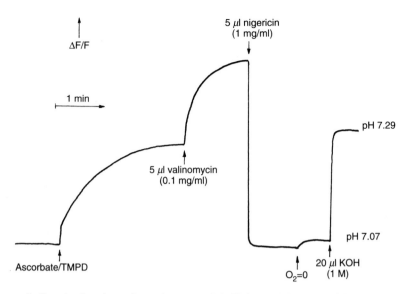

Figure 2. Respiration-dependent changes of ΔpH in cytochrome oxidase proteolipo-somes. Changes in the internal pH of cytochrome oxidase proteoliposomes were moni-tored by fluorescence changes of the entrapped pH probe pyranine. The proteoliposomes were prepared by sonication according to *Protocol 1*. The 3 ml cuvette contained proteo-liposomes (5 mg asolectin, 0.2 nmol cytochrome *c* oxidase) with 5 mM entrapped pyra-mine in 0.1 m K^+-HEPES, originally at pH 7.2. The vesicles were suspended in 0.1 M K^+-HEPES pH 7.0, and allowed to equilibrate for 10 min at 30°C with continuous stirring. Cytochrome *c* was added to a final concentration of 1.7 μM followed by a mixture of ascorbate (1.7 mM) plus TMPD (50 μM). Fluorescence changes were monitored using a long-pass cut-off filter (cut-off at 480 nm) with excitation at 460 nm. The addition of reductant initiates enzyme activity and fluorescence increases as the interior of the vesicles becomes alkaline. The addition of valinomycin abolishes any respiration-dependent Δψ and, as a result, the passive influx of H^+ into the vesicles is lowered. The consequence is that the system can now sustain a higher steady state level of ΔpH and the fluorescence increases further (39). Addition of the electroneutral ionophore nigericin equilibrates H^+ and K^+ gradients and ΔpH collapses. A slight changes in fluorescence occurs on anaerobiosis due to changes in the quenching effect of cytochrome *c*. Calibra-tion of internal fluorescence changes with pH can be made by the addition of KOH.

4.3 Measurement of Δψ

The basis of most Δψ measurements is the use of probe molecules that are lipophilic enough to freely permeate the membrane but are also charged so that they distribute according to the Nernst equation:

$$\Delta\psi \,(\text{mV}) = 59 \log \frac{[\text{ion}]_{\text{in}}}{[\text{ion}]_{\text{out}}}.$$

The distribution of the ionic probe under the influence of a Δψ can then be monitored in a number of ways depending on the molecular properties of the probe (see Chapter 3). Radiotracer methods can be used, for example Rb^+ to measure $\Delta\psi$ in cytochrome oxidase proteoliposomes (47), but for many bioenergetic studies a continuous monitor where short time kinetic measurements can be made is an advantage. This can be provided using ion-specific electrodes to monitor the concentrations of lipophilic cations such as butyltriphenyl phosphonium ($BTPP^+$) or tetraphenylphosphonium (TPP^+), and also lipophilic anions such as phenyldicarbaundecaborane (PCB^-) (48, 49). The electrodes themselves are not difficult to make (see refs 50, 51, and Chapter 3). Their response to changes in ion concentration is logarithmic and calibration with standard concentrations of the specific ion is required. As mentioned above in relation to the measurements of ΔpH by probe distribution, the trapped volume of the vesicles is required to calculate the internal concentration of the probe. The electrode traces can then be converted to changes in $\Delta\psi$. High (> 1 mM) concentrations of probe should be avoided, if sensitivity allows, since the probes themselves can act to neutralize $\Delta\psi$. Optical probes are used at lower (μM) concentrations and thus perturb the system much less. These usually depend on absorbance or fluorescence quenching to monitor internal concentration changes. Cyanine and safranine dyes are examples of $\Delta\psi$ probes that accumulate in cells or vesicles where the interior potential is negative (52, 53). Accumulation results in 'stacking' which causes optical interaction between the molecules. The response of the cyanine dye 3',3'-dipropylthiadicarbocyanine ($diSC_3$-5) to $\Delta\psi$ in an active, energized system is shown in *Figure 3*.

Unfortunately the mechanism of action of most $\Delta\psi$ probes involves interaction with the interior surface of the membrane. Accumulation of probe occurs on membrane binding sites. This means that the probe/lipid ratio is an important factor governing the response of the system to $\Delta\psi$ (54). Nonlinearity can occur at high $\Delta\psi$ values due to saturation of the available binding sites, and also at low $\Delta\psi$ values where interaction between the chromophores is low. Ionic strength can also affect dye response as can the presence of fluorescence quenching molecules such as cytochrome c (55). Initial experiments should therefore be done in any investigation to determine the linearity of response of the probe with vesicle concentration, otherwise quantitative comparisons between different systems is not possible. Where measurements are to be made on vesicles with a positive internal potential, anionic probes can be used. The oxonol series of dyes (56) have been shown to respond to such potentials in proteoliposomes incorporating H^+-ATPase (57), Na^+/K^+-ATPase (58), and cytochrome oxidase (59).

It is not possible to relate the optical changes in dye concentration directly to $\Delta\psi$. Hence it is necessary to calibrate the changes in absorbance or fluorescence using artificial diffusion potentials of known magnitude. For the anionic dyes, valinomycin plus the addition of a known concentration of K^+ can induce a known potential (positive inside the vesicle) if the vesicles are originally made in Na^+ medium containing a small but known concentration

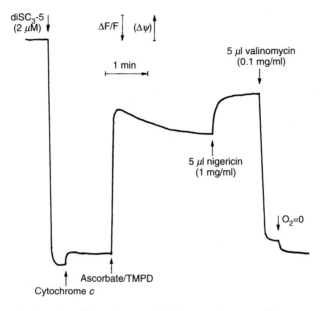

Figure 3. Respiration-dependent changes of $\Delta\psi$ in cytochrome oxidase proteoliposomes. Changes in $\Delta\psi$ of cytochrome oxidase proteoliposomes were monitored by fluorescence changes of the positively charged lipophilic probe diSC$_3$-5 (see section 4.3). The proteoliposomes were prepared by sonication according to *Protocol 1*, and the conditions of measurement were as described in the legend to *Figure 2*. Fluorescence changes of disSC$_3$-5 were monitored using a long-pass cut-off filter at 670 nm (excitation at 622 nm). The addition of diSC$_3$-5 causes a rise in fluorescence (in the above trace, fluorescence increase is in the downward direction). When the reductant ascorbate (plus TMPD) is added, cytochrome c becomes reduced and initiates enzyme activity. Proton pumping by the oxidase immediately establishes a $\Delta\psi$ across the vesicle membrane. The resulting redistribution of the probe to the vesicle interior causes self-quenching and the fluorescence intensity falls (a fast rise in the trace). Slow changes in $\Delta\psi$ occur as ΔpH is established (see (39)). The addition of nigericin abolishes ΔpH and all the protonmotive force is now expressed in an increased $\Delta\psi$. Valinomycin addition allows the electrogenic movement of K$^+$ to abolish $\Delta\psi$ and the fluorescence intensity rises back to its original level.

of K$^+$ (59). For potentials of the opposite polarity, the proteoliposomes have to be loaded internally with K$^+$. For example, Singh and Nicholls (54) have prepared proteoliposomes of cytochrome oxidase by sonication using a medium of 20 mM K$^+$-Hepes buffer in 100 mM KCl (pH 7.4). Untrapped K$^+$ was removed by Sephadex G-50 chromatography in Na$^+$ medium and diffusion potentials generated by diluting the vesicles into a medium containing different concentrations of K$^+$. The K$^+$ and Na$^+$ concentrations had to be adjusted to maintain constant ionic strength. With this technique, it has been possible to calibrate cationic dye response up to values of 180 mV. The same workers have compared $\Delta\psi$ calibrations between the optical probes safranine and diSC$_3$-5 and the specific-ion electrode probe BTPP$^+$ using cytochrome oxidase proteoliposomes (51). They found that $\Delta\psi$ estimated by BTPP$^+$ distribution was lower than that calculated by the optical probes, around 80 mV compared to values of around 150 mV. They attribute these discrepancies to be due in part to an overestimation of the imposed diffusion potentials when the K$^+$ loading method is used to calibrate the fluorescence probes (efflux of the internal K$^+$ will occur during separation of the vesicles from the external K$^+$ medium by chromatography) and an underestimation of the potential when the ion-specific electrode method is used because of partial neutralization of $\Delta\psi$ by the permeant probe itself and possible non-linear binding of the probe at high $\Delta\psi$. Similar discrepancies have also been noted (60) between the values of $\Delta\psi$ measured by the ion-selective electrode method (TPP$^+$) and by changes in carotenoid absorbance (see below).

The electrochromic response of an intrinsic probe such as provided by carotenoids can be used to monitor transmembrane potentials. These molecules show a $\Delta\psi$-dependent shift in their absorbance spectrum, which is linear with imposed diffusion potentials (61) and responds rapidly to changes in $\Delta\psi$. Active pigment complexes from a *Rhodobacter sphaeroides* have been reconstituted into liposomes and have shown similar responses to $\Delta\psi$ as in the native membrane (62, 63). Since the direction of the absorbance change in the native membrance is dependent on the polarity, this implies a unidirectional orientation of the pigment complexes in the proteoliposome membrane. The disadvantage of the system is that the absorbance changes can be complicated by spectral interference from other compounds in the protein complexes. Nevertheless, the electrochromic response of incorporated carotenoid/ protein complexes could be useful for $\Delta\psi$ measurements in proteoliposomes co-reconstituted with ion-translocating proteins.

5. Double reconstitutions

Double reconstitution provides one of the most powerful techniques for studying the interaction of bioenergetic systems. It was the historical double reconstitution experiments of Racker and colleagues (64) that helped to provide a firm experimental basis for Mitchell's chemiosmotic theory. However, there are several problems with double reconstitutions. Membrane proteins vary in their sensitivity to the different reconstitution procedures. A compromise may have to be made where activity loss in one protein during reconstitution is traded against more efficient incorporation of the second protein. A further difficulty is that the orientation of the two proteins in the bilayer has to be such that both interact in a meaningful way. For example, co-reconstitution of cytochrome oxidase with purified mitochondrial ATPase by detergent dialysis is difficult because the oxidase mainly orients itself in the membrane so as to reduce externally added cytochrome *c*. To couple the oxidase system to ATP synthesis requires the ATPase to incorporate with its

4: Reconstitution of bioenergetic proteins and uses of proteoliposomes

ATP binding site on the inside of the vesicle. Unfortunately, the reconstituted H^+-ATPase mainly orients in the opposite direction by this reconstitution procedure. Thus the two enzymes are not coupled by a common H^+ gradient in the reconstituted system. A small population of vesicles will co-reconstitute with the correct orientations for coupling but these appear to be few in number, giving a low coupling efficiency overall (65). One approach for these two enzymes could be to reconstitute by sonication where a more random incorporation of cytochrome oxidase occurs. Entrapped cytochrome c could be used as substrate (see Cooper and Nicholls (59) for example). The problem then would be the low reconstitution efficiency of the H^+-ATPase when the sonication method is used (66).

Some systems naturally co-reconstitute with the necessary orientation for vectorial coupling. Reconstitution of bacteriorhodopsin by the detergent dialysis method yields proteoliposomes that pump protons to the exterior on illumination. When co-reconstituted with H^+-ATPase from mitochondria (64), chloroplasts (67), bacteria (68), or plasma membrane (66) (see *Protocol 4*), net ATP synthesis can be shown by light-induced H^+ gradients.

Protocol 4. Co-reconstitution of yeast plasma membrane H^+-ATPase and bacteriorhodopsin (66)

Equipment and reagents

- Solid asolectin washed according to Darley-Usmar et al. (10)
- Purified bacteriorhodopsin from *Halobacterium halobium* (69) solubilized in 80 mM octyl glucoside at a protein concentration of around 2 mg/ml
- Purified plasma membrane H^+-ATPase

from yeast (70) (suspended in 0.1 M Tris–HCl pH 7.0, 25 mM K_2SO_4, 1mM $MgSO_4$, 0.5 M sucrose, at a protein concentration of approx. 0.3 mg/ml)
- 25 mM K_2SO_4, 1 mM $MgSO_4$, 0.1 M Tris–HCl pH 7.0
- Octyl glucoside

Method

1. Suspend 100 mg asolectin in 10 ml 25 mM K_2SO_4, 1 mM $MgSO_4$, 0.1 M Tris–HCl pH 7.0 and vortex for 3 min. Sonicate to clarity in a bath-type sonicator.

2. Add octyl glucoside to a final concentration of 45 mM followed by the H^+-ATPase (final concentration of 400 nM) and the bacteriorhodopsin (final concentration 5 µM).

3. Dialyse in the dark for 48 h at 4°C against a total of 2.5 litres of 25 mM K_2SO_4, 1 mM $MgSO_4$, 0.1 M Tris-HCl pH 7.0, with several changes of buffer (see *Protocol 2*).

It is not always necessary to purify both protein components for a co-reconstitution study. Fusing proteoliposomes incorporating one of the proteins with a fraction of membrane vesicles containing the other protein, still in its biological *milieu*, can often provide a useful system for study. For example, membrane vesicles derived from *Streptococcus cremoris* have been fused with proteoliposomes incorporating cytochrome oxidase from beef heart (71). The resulting vesicles were used to study the effect of imposed proton motive force on the secondary transport of amino acids. Similarly, galactose transport has been studied in a membrane preparation formed by fusing cytochrome oxidase proteoliposomes with plasma membranes from the yeast *Kluyveromyces marxianus* (see *Protocol 5* and ref. 19). The effects of protonmotive force on the galactose transporter could then be measured using cytochrome oxidase to impose a pH gradient and membrane potential.

Protocol 5. Fusion of cytochrome oxidase proteoliposomes with yeast plasma membranes (19)

Equipment and reagents

- Cytochrome oxidase proteoliposomes prepared by detergent dialysis as in *Protocol 2* but using 0.75% octyl glucoside instead of cholate
- Plasma membrane vesicles purified from yeast according to Van Leeuwen *et al.* (19)

Method

1. Mix the proteoliposomes and plasma membrane vesicles together in a plasma membrane protein:proteoliposome lipid ratio of 1:20 in 50 mM potassium phosphate buffer pH 6.2, containing 1 mM MgCl$_2$ to a volume of 300 µl in a 1.5 ml microcentrifuge tube.
2. Freeze in liquid nitrogen and thaw at room temperature.
3. Sonicate for two 10 sec bursts using a sonication bath. The suspension should become clear.

References

1. Casey, R. P. (1984). *Biochim. Biophys. Acta*, **768**, 319.
2. Villalobo, A. (1990). *Biochim. Biophys. Acta*, **1017**, 1.
3. New, R. R. C. (1990). *Liposomes: a practical approach*. IRL Press, Oxford.
4. Allen, T. M., Romans, A. Y., Kercret, H., and Segrest, J. P. (1980). *Biochim. Biophys. Acta*, **601**, 328.
5. Wrigglesworth, J. M. (1985). In *Structure and properties of cell membranes* (ed. G. Benga), Vol. 1, pp. 137–58. CRC Press Inc., Boca Raton, Florida.
6. Robinson, N. C. (1982). *Biochemistry*, **21**, 184.
7. Huganir, R. and Racker, E. (1982). *J. Biol. Chem.*, **257**, 9372.
8. Madden, T. D. and Cullis, P. R. (1985). *Biochim. Biophys. Acta*, **808**, 219.
9. Cooper, C. E., Wrigglesworth, J. M., and Nicholls, P. (1990). *Biochem. Biophys. Res. Commun.*, **173**, 1008.

10. Darley-Usmar, V. M., Capaldi, R. A., Takamiya, S., Millett, F., Wilson, M. T., Malatesta, F., and Sarti, P. (1987). In *Mitochondria: a practical approach* (ed. V. M. Darley-Usmar, D. Rickwood, and M. T. Wilson), pp. 113–29. IRL Press, Oxford.
11. Madden, T., Hope, M., and Cullis, P. (1984). *Biochemistry*, **23**, 1413.
12. Tihova, M., Tattrie, B., and Nicholls, P. (1993). *Biochem. J.*, **292**, 933.
13. Chicken, C. A. and Sharom, F. J. (1984). *Biochim. Biophys. Acta*, **774**, 110.
14. Ketis, N. V. and Grant, C. W. M. (1983). *Biochim. Biophys. Acta*, **730**, 359.
15. Wrigglesworth, J. M., Wooster, M. S., Elsden, J., and Danneel, H.-J. (1987). *Biochem. J.*, **246**, 737.
16. Racker, E., Violand, B., O'Neal, S., Alfonzo, M., and Telford, J. (1979). *Arch. Biochem. Biophys.*, **198**, 470.
17. Kasahara, M. and Hinkle, P. C. (1977). *J. Biol. Chem.*, **252**, 7384.
18. Pick, U. (1981). *Arch. Biochem. Biophys.*, **212**, 186.
19. Van Leeuwen, C. C. M., Postma, E., Vanden Broek, P. J. A., and Van Steveninck, J. (1991). *J. Biol. Chem.*, **266**, 12146.
20. Olson, F., Hunt, C. A., Szoka, F. C., Vail, W. J., and Papahadjopoulos, D. (1979). *Biochim. Biophys. Acta*, **557**, 9.
21. Hope, M. J., Bally, M. B., Webb, G., and Cullis, P. R. (1985). *Biochim. Biophys. Acta*, **812**, 55.
22. Madden, T. D. and Cullis, P. R. (1984). *J. Biol. Chem.*, **259**, 7655.
23. Dencher, N. A. (1986). *Biochemistry*, **25**, 1195.
24. Driessen, A. J. M., Hellingwerf, K. J., and Konings, W. N. (1985). *Biochim. Biophys. Acta*, **808**, 1.
25. Gale, P. and Watts, A. (1991). *Biochem. Biophys. Res. Commun.*, **180**, 939.
26. Wrigglesworth, J. M. (1983). In *Biochemical research techniques* (ed. J. M. Wrigglesworth), pp. 147–76. John Wiley & Sons Ltd.
27. Wrigglesworth, J. M. (1985). *J. Inorg. Biochem.*, **23**, 311.
28. Bayerl, T. M., Schmidt, C. F., and Sackmann, E. (1988). *Biochemistry*, **27**, 6078.
29. Charbonnier, M., Lechene de la Porte, P., Veesler, S., Pike, E. R., Woodley, W. A., and Patin, P. (1993). *Int. Labmate*, **18**, 37.
30. Barenholtz, Y., Gibbes, D., Litman, B. J., Goll, J., Thompson, T. E., and Carlson, F. D. (1977). *Biochemistry*, **16**, 2806.
31. Popot, J.-L., Cartaud, J., and Changeux, J.-P. (1981). *Eur. J. Biochem.*, **118**, 203.
32. Lind, C., Hojjeberg, B., and Khorana, H. G. (1981). *J. Biol. Chem.*, **256**, 8298.
33. Driessen, A. J. M., De Vrij, W., and Konings, W. N. (1985). *Proc. Natl. Acad. Sci. USA*, **82**, 7555.
34. Wrigglesworth, J. M. (1978). *FEBS Symp.*, **45**, 95.
35. Nicholls, J. M., Hildebrandt, V., and Wrigglesworth, J. M. (1980). *Arch. Biochem. Biophys.*, **204**, 533.
36. Zhang, Y.-Z., Capaldi, R. A., Cullis, P. R., and Madden, T. D. (1985). *Biochim. Biophys. Acta*, **808**, 209.
37. Smythe, C. P. (1955). *Dielectric behaviour and structure*. McGraw-Hill, New York, Toronto.
38. Pauly, H., Packer, L., and Schwan, H. P. (1960). *J. Biochem. Biophys. Cytol.*, **7**, 589.
39. Wrigglesworth, J. M., Cooper, C. E., Sharpe, M. A., and Nicholls, P. (1990). *Biochem. J.*, **270**, 109.

40. Hinkle, P. C. (1973). *Fed. Proc. Fed. Am. Soc. Exp. Biol.*, **32**, 1988.
41. Azzone, G. F., Pietrobon, D., and Zoratti, M. (1984). *Curr. Top. Bioenerg.*, **13**, 1.
42. Harrigan, P. R., Hope, M. J., Redelmeier, T. E., and Cullis, P. R. (1992). *Biophys. J.*, **63**, 1336.
43. Rudelmeier, T. E., Mayer, L. D., Wong, K. F., Bally, M. B., and Cullis, P. R. (1989). *Biophys. J.*, **56**, 385.
44. Cafiso, D. S. and Hubbell, W. L. (1978). *Biochemistry*, **17**, 3871.
45. Nichols, J. W., Hill, M. W., Bangham, A. D., and Deamer, D. W. (1980). *Biochim. Biophys. Acta*, **596**, 393.
46. Hellingwerf, K. J. and Van Hoorn, P. (1985). *J. Biochem. Biophys. Methods*, **11**, 83.
47. Krishnamoorthy, G. and Hinkle, P. C. (1984). *Biochemistry*, **23**, 1640.
48. Jasaitis, A. A., Nemecek, I. B., Skulachev, V. P., and Smirnova, S. M. (1972). *Biochim. Biophys. Acta*, **275**, 485.
49. Drachev, L. A., Jasaitis, A. A., Kaulen, A. D., Kondrashin, A. A., Chu, L. V., Semenov, A. Y., Severina, I. I., and Skulachev, V. P. (1976). *J. Biol. Chem.*, **251**, 7072.
50. Kamo, N., Muratsugu, M., Hongoh, R., and Kobatake, Y. (1979). *J. Membr. Biol.*, **49**, 105.
51. Singh, A. P. and Nicholls, P. (1986). *Arch. Biochem. Biophys.*, **245**, 436.
52. Waggoner, A. S. (1979). *Annu. Rev. Biophys. Bioeng.*, **8**, 47.
53. Azzone, G. F., Pietrobon, D., and Zoratti, M. (1984). *Curr. Top. Bioenerg.*, **13**, 1.
54. Singh, A. P. and Nicholls, P. (1985). *J. Biochem. Biophys. Methods*, **11**, 95.
55. Singh, A. P., Chanady, G. A., and Nicholls, P. (1985). *J. Membr. Biol.*, **84**, 183.
56. Bashford, C. L., Chance, B., and Prince, R. C. (1979). *Biochim. Biophys. Acta*, **545**, 46.
57. Walraven, H. S., Hagendoorn, M. J. M., Krab, K., Haak, N. P., and Kraayenhof, R. (1985).
58. Appel, H.-J. and Berrsch, B. (1987). *Biochim. Biophys. Acta*, **903**, 480.
59. Cooper, C. E. and Nicholls, P. (1990). *Biochemistry*, **29**, 3859.
60. Crielaard, W., Cotton, N. P. J., Jackson, J. B., Hellingwerf, K. J., and Konings, W. N. (1988). *Biochim. Biophys. Acta*, **932**, 17.
61. Jackson, J. B. and Crofts, A. R. (1969). *FEBS Lett.*, **4**, 185.
62. Crielaard, W., Helling, K. J., and Konings, W. N. (1989). *Biochim. Biophys. Acta*, **973**, 205.
63. Goodwwin, M. G. and Jackson, J. B. (1993). *Biochim. Biophys. Acta*, **1144**, 191.
64. Racker, E. and Stoeckenius, W. (1974). *J. Biol. Chem.*, **249**, 662.
65. Racker, E. and Kandrach, A. (1971). *J. Biol. Chem.*, **246**, 7069.
66. Wach, A., Dencher, N. A., and Graber, P. (1993). *Eur. J. Biochem.*, **214**, 563.
67. Richard, P. and Graber, P. (1992). *Eur. J. Biochem.*, **210**, 281.
68. Sone, N., Takeuchi, Y., Yoshida, M., and Ohno, K. (1977). *J. Biochem.*, **82**, 1751.
69. Bauer, P. J., Dencher, T. N. A., and Heyn, M. P. (1976). *Biophys. Struct. Mech.*, **2**, 79.
70. Wach, A., Ahlers, J., and Graber, P. (1990). *Eur. J. Biochem.*, **189**, 675.
71. Driessen, A. J. M., De Vrij, W., and Konings, W. N. (1985). *Proc. Natl. Acad. Sci. USA*, **82**, 7555.

5

Redox states and potentials

RICHARD CAMMACK

1. Introduction

This chapter describes methods for controlling the oxidation-reduction, or redox, state of electron transfer proteins, or membrane bioenergetic systems. There are two principal objectives:

(a) To fix, or poise, the redox centres (e.g. haem, iron-sulfur clusters, copper, molybdenum, flavins, quinones) in defined redox states, in order to study their biochemical, chemical, or spectroscopic properties.

(b) To measure midpoint reduction potentials, also known as redox potentials $(E^{0'})$ of redox components. These are important properties, since they define the driving force for electron transfer processes.

1.1 Methods of oxidation and reduction of samples

The redox potential of the system may be adjusted by:

(a) Specific enzyme substrates, such as succinate or NAD(P)H.

(b) Non-specific chemical reductants such as dithionite or ascorbate, or oxidants such as ferricyanide.

(c) Photochemical oxidation or reduction. In this way extreme redox potentials may be achieved. For non-photosynthetic systems, photosensitizers such as flavins are used.

(d) Electrochemically at a working electrode of gold, platinum, or carbon, using a potentiostat (1–3).

This article will concentrate on chemical methods of oxidation and reduction, but first a number of alternative methods will be considered.

1.2 Methods for estimation of redox potentials

1.2.1 Direct electrochemical methods

It is possible to reduce or oxidize samples by electrolysis. Electrochemical reduction is slower in approaching equilibrium, but it avoids the use of reactive chemicals such as dithionite. It is also possible to measure redox potentials directly by voltammetry or coulometry. Cyclic voltammetry and differential pulse polarography are dynamic methods, which are rapid and require little material. The potential of an electrode is swept, and a 'wave' in the current/voltage plot is observed which corresponds to the potential of a reducible species. These are absolute methods and do not require any spectroscopic observations. With careful controls they can be made quantitative. The interaction of the proteins with the electrode is assisted by *promoters*, small molecules with suitable charge or hydrophobicity properties which do not directly participate in the redox reaction. The appropriate promoters must be found by trial and error, and this may present difficulties.

Voltammetric methods have been applied extensively to small molecules and to some purified proteins (4). For the complex redox enzymes and membrane bound complexes found in bioenergetic systems, which contain several redox centres, and other reducible groups such as disulfides, it is more difficult to assign the redox potentials obtained from voltammetry to individual redox centres. However some interesting dynamic effects have been observed (5).

1.2.2 Equilibration with a standard

The redox system may be allowed to come to equilibrium with a redox couple of known potential. The redox potential may then be calculated from the Nernst equation (2). Suitable redox couples include:

(a) Redox dyes, where the extent of reduction may be observed spectrophotometrically.

(b) Mixtures of oxidized and reduced substrates, such as NAD^+ and NADH, for appropriate enzymes.

(c) Partial pressures of hydrogen in mixtures with inert gas, in the presence of catalytic concentrations of hydrogenase and methyl viologen as a mediator.

The useful potential range of these couples depends on their midpoint potentials, and also on pH (see *Table 1*).

1.2.3 Spectroelectrochemistry

A method, which will be described here in detail, combines potential measurements and spectroscopy. A protein or membrane bound complex is equilibrated at a particular potential by addition of non-specific oxidizing or reducing agents. The redox potential is measured by means of an electrode, while the redox state of the protein is determined spectroscopically. This provides the microscopic redox potentials of any redox centre in the protein which can be detected, in either its oxidized or reduced form. The spectroscopic methods may be divided into:

(a) Methods where measurements can be made continuously on the solution. The principal method is optical absorption, which is particularly effective for cytochromes. Circular dichroism is less sensitive, and has been less frequently used, but has the advantage that the colours of mediators do not interfere (6).

(b) Methods where samples have to be taken for measurements, for example because they are made in the frozen state. These include electron paramagnetic resonance (EPR), which is useful for free radicals, and for iron-sulfur clusters and other transition metal complexes. The protocols described here are specifically for this method. However the methods may be adapted to prepare samples for Mössbauer spectroscopy of iron compounds, and magnetic circular dichroism of chromophoric transition metal proteins.

Examples of both types of method will be described.

Although it is possible in favourable cases to achieve equilibrium directly between the protein(s) and an electrode, this is generally slow. The process is facilitated by the use of *mediators*, which are compounds, such as dyes. Unlike promoters, mediators themselves participate in oxidation–reduction. They are selected to have redox potentials in the range to be studied (see *Table 1*). They communicate the redox potential of the solution to the electrode, and also serve as a redox 'buffer' of the system. The appropriate choice of

Table 1. Properties of some redox couples

Compound	E^0, mV versus S.H.E., pH 7	No. of electrons transferred
Potassium ferricyanide	+420	1
1,2-benzoquinone	+372	2
2,6,2'-trichloroindophenol	+254	2
Diaminodurol	+240	2
2,6-dichloroindophenol	+217	2
Phenazine methosulfate	+80	1–2
Thionine	+60	2
Phenazine ethosulfate	+55	1–2
Duroquinone	+7	2
Methylene blue	−11	1–2
Pyocyanin	−60	2
Resorufin	−50	2
Indigotrisulfonate	−81	2
Indigodisulfonate	−125	2
Anthraquinone-1,5-disulfonate	−170	2
Anthraquinone-2,6-disulfonate	−184	2
Anthraquinone-2-sulfonate	−225	2
Phenosafranin	−255	2
Safranin T	−289	2
Benzyl viologen	−360	1
NAD$^+$/NADH	−320	2
Methyl viologen	−440	1
H$^+$/H$_2$	−414	2

mediators, to be effective without interfering with the measurements, is an important consideration in these experiments.

2. Techniques and apparatus

Many types of apparatus have been constructed for oxidation–reduction of biological samples and clearly there is no single best way to do this. This article will describe a relatively simple and flexible apparatus which can be adapted to a number of different situations. This apparatus requires some glass construction, but various types of glassware can be adapted for these purposes, so there will be an emphasis on the general principles involved.

2.1 Maintenance of oxygen-free conditions

It must be accepted that in any apparatus which is not hermetically sealed, oxygen will inevitably leak in. The objective of the techniques is to keep this oxygen down to an acceptable level. As far as possible, the apparatus should be of glass and connected by glass or metal tubing with tightly sealed joints. Rubber and plastics are permeable to oxygen, some more so than others. Butyl rubber, for example, is much less permeable than silicone. For rigorous oxygen-free conditions it may be necessary to use a vacuum line or an anaerobic glove box with a recirculating purified nitrogen atmosphere. However if a pure gas line is available, an apparatus such as that in *Figure 1* may be used.

A supply of pure inert gas, either nitrogen or argon, is needed. Argon is preferable, though more expensive, because of its higher purity. Its higher density also means that it mixes more slowly with air in open tubes. It is preferable to start with pure gas, and prevent oxygen getting in, than to try

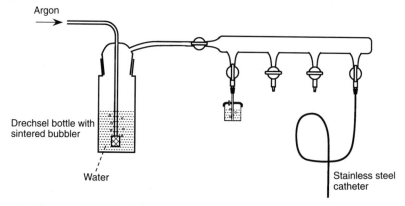

Figure 1. Glass manifold for distribution of inert gas (nitrogen or argon). The outlets are vacuum stopcocks fitted with Luer tips from glass syringes. The gas bubbler contains water, to prevent drying out of samples.

to remove oxygen afterwards. Some authors have suggested various ways of scrubbing oxygen from cylinder gases by bubbling through reducing solutions. Generally this is ineffective as removal of oxygen traces through a gas–liquid interface is very slow, even with efficient agitation. The reducing solutions also tend to form an aerosol in the gas stream, which will find its way into the sample. It is possible to remove traces of oxygen from inert gases by passing them through a heated column packed with copper or BASF catalyst. However if precautions are taken to prevent oxygen leakage, cylinder gases such as oxygen-free nitrogen or argon are usually sufficiently pure. A pressure regulator with a rubber diaphragm is a significant source of oxygen contamination. A special regulator with a stainless steel diaphragm is essential.

Gas lines should be checked for traces of oxygen. A simple glass apparatus for testing the presence of oxygen, which uses photochemically reduced methyl viologen as indicator, has been described (7). Fuel cell oxygen monitors may be obtained (Systek, Oxanal), which measure oxygen levels down to one to five parts per million. Apparatus may be checked for leaks by using indicator solutions, such as reduced methyl viologen which changes from blue to colourless on oxidation.

Long syringe needles, 0.9 mm diameter (20 gauge), up to one metre long, with Luer fittings, may be specially ordered from medical suppliers. They are attached to a gas manifold consisting of a series of glass syringes and stopcocks (see *Figure 1*) which in turn is connected to the gas cylinder. This type of apparatus may be adapted for cells for electrochemical measurements, in a spectrophotometer.

Solutions may be made anaerobic either by bubbling with inert gas, or by repeatedly evacuating with a vacuum pump and flushing with inert gas. For solutions containing proteins or detergents, frothing may occur, leading to denaturation. Alternatively, the inert gas may be flowed over the surface; equilibration between the gas and liquid phases is slow, but is assisted by vigorous stirring. If the surface is small it may take hours to decrease the oxygen in solution to an acceptable level. However it should be remembered that dithionite and many reduced mediators will scavenge oxygen during reductive titration.

2.2 Oxidizing and reducing agents

Sodium dithionite (sodium hydrosulfite), $Na_2S_2O_4$, is an effective general purpose reducing agent. The limit of the redox potential achieved in practice is about -470 mV at pH 7. Lower potentials are possible at more alkaline pH, decreasing by 59 mV/pH unit. Commercial dithionite contains various impurities, mainly sodium bisulfite and sodium carbonate. These do not usually interfere with its action as a general reducing agent, but for quantitative work, it may be purified by crystallization (8). Dithionite solutions react immediately with oxygen and must be kept under an inert gas atmosphere in a septum-stoppered vessel. Dithionite forms acid bisulfite on oxidation, so the solutions must be adequately buffered. Alkaline solutions are stable for a few hours. Other reducing agents include ascorbate, and substrates for enzymes if appropriate. Potassium ferricyanide, $K_3Fe(CN)_6$, is a good general purpose oxidizing agent. Solutions are stable in the dark, but release cyanide in the light. It can achieve potentials up to about +400 mV. In excess concentrations ferricyanide can react with some proteins such as iron–sulfur proteins, changing [4Fe–4S] clusters into [3Fe–4S] (9).

Flavins, such as riboflavin, are strong reducing agents in blue light. A typical concentration used is 5–10 μM riboflavin. Deazaflavin or its more water soluble derivatives are better, as they do not yield free radicals. The solution, in a glass or quartz vessel, is irradiated with a quartz–iodine lamp. A sacrificial electron donor must be present in the solution; 5 mM EDTA is commonly used, though we have found that 'Good' buffers such as Hepes will also work.

The simplest type of experiment is where a reducing agent is added in excess to a biochemical preparation in a sample container such as an EPR tube (*Protocol 1*). Any residual traces of oxygen in the solution are removed by strong reducing agents (such as dithionite), or by enzymes in the sample (such as cytochrome *c* oxidase). Anaerobic conditions are maintained by a flow of inert gas (nitrogen or argon) over the surface of the solution. Finally, the sample is frozen.

Mitochondria may be obtained in different redox states, by adding substrates with reduction potentials sufficient to reduce particular redox components. The method described here uses succinate as the reducing agent. The E^0 of the fumarate/succinate couple is +28 mV, so succinate is able to reduce cytochromes *c*, *a*, and to a certain extent *b*, the [2Fe–2S] and [3Fe–4S] clusters of Complex II and the Rieske [2Fe–2S] iron–sulfur cluster of Complex III, but not most of the other iron–sulfur clusters. When succinate is added to mitochondria, as soon as all oxygen in the solution is reduced by cytochrome *c* oxidase, any remaining endogenous NAD-linked substrates will further reduce the respiratory chain. To avoid this it is necessary to use well-washed mitochondria, or, as described here, submitochondrial particles. Since succinate dehydrogenase tends to be inactive in the resting state, due to tight binding of the inhibitor oxaloacetate, an activating step is used first.

Protocol 1. Preparation of a sample of submitochondrial particles for EPR spectroscopy

Equipment and reagents

- Submitochondrial particles, concentrated by centrifugation to a suitable concentration for EPR spectroscopy (at least 20 mg/ml)
- Supply of inert gas, supplied through butyl rubber tubing, terminating in a syringe needle

(about 0.8 mm diameter), long enough to reach nearly to the bottom of the sample tube
- Reductant (500 mM sodium succinate stock solution) stored in a septum-sealed vial flushed with inert gas

- Activating solutions for succinate dehydrogenase: 100 mM MgCl₂, 10 mM ATP, 100 mM KH₂PO₄
- Microlitre syringes to transfer solutions (these should have needles long enough to reach the bottom of the EPR tube)
- Quartz EPR sample tubes (Wilmad Glass Co.)
- Freezing bath: methanol, in a polystyrene coffee cup, cooled in liquid nitrogen—stir until the liquid becomes slightly viscous

Method

1. Transfer 150 µl submitochondrial suspension into the bottom of the EPR tube with a syringe or Pasteur pipette. Shake the tube, like a clinical thermometer, to drive the liquid to the bottom if necessary.

2. The gas flow through the needle should be strong enough to prevent oxygen from entering the EPR tube (check by bubbling through water). Flush the EPR sample tube for 30 sec, being careful not to put the gassing needle below the surface of the solution, which would drive it out of the tube.

3. To activate succinate dehydrogenase, add to the membranes 5 mM MgCl₂, 1 µM/mg protein ATP, and 1 µM/mg protein KH₂PO₄, and leave to stand for 10 min at room temperature, under inert gas.

4. Inject 10 µl of 500 mM succinate into the submitochondrial suspension, and mix by vigorously stirring up and down with the syringe needle for 2 min, while maintaining the inert gas flow.

5. Allow the solution to stand for 2 min.

6. Immerse the tube into the freezing bath. Alternatively, cautiously immerse it into liquid nitrogen.

7. Store the sample over liquid nitrogen.

2.3 Methods of oxidation–reduction of samples

For oxidation and reduction of samples it is necessary to make precise additions of solutions under anaerobic conditions. The use of an anaerobic glove box is effective but cumbersome. For procedures such as stoichiometric titrations, or redox potential measurements, it is possible to construct relatively simple glassware. Vacuum line techniques are well developed for the chemistry of oxygen-sensitive substances. The whole apparatus must be made vacuum-tight with glass stopcocks and glass conical joints (10, 11). These are somewhat fragile. Biochemical work entails the use of aqueous solutions, often in very small volumes, so drying out under vacuum is a problem.

The approach developed in our laboratory has been to use only a slight positive pressure of an inert gas. This is distributed through a glass manifold, as used in the vacuum line systems, through fine stainless steel tubing or syringe needles. Drying down or concentration of solutions may be minimized by bubbling the gas through water first (at the same temperature as the solution). As long as a continuous flow of gas is maintained it is possible to use rubber septa, and arrange the apparatus so that any oxygen which leaks in is flushed away.

In this apparatus, solutions are transferred with gas-tight microlitre syringes, and for sensitive work the apparatus is arranged so that no part of the solution is exposed to the air.

Syringes and gas lines are connected to the apparatus through small rubber septa, which also allow some flexibility in transferring solutions from one part of the apparatus to another. For syringes which have to be moved in and out frequently, the septa are penetrated with small stubs of needle tubing just wide enough (16 gauge, 1.65 mm outer diameter), for the syringe needles to be slid through. These stubs of tubing also act as outlets for the inert gas, so that any oxygen that leaks in, tends to be flushed away.

More rigorous oxygen-free conditions are needed for stoichiometric titrations, for example determining the number of reducing equivalents required to reduce a protein. In this case, the solution of reducing agent must be standardized. This may be done spectrophotometrically, in a cell such as that shown in *Figure 2*. Dithionite solution may be standardized in an anaerobic spectrophotometer cuvette by titrating into oxygen-free buffer. The molar absorbance at 314 nm is 8.0/M/cm (12).

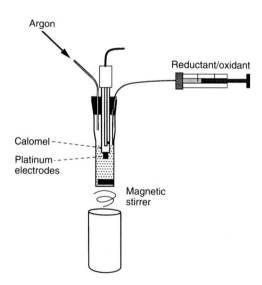

Figure 2. A cell for controlled-potential titration of protein solutions. The lower part is a glass spectrophotometer cuvette, fused on to a neck with conical section. The combined platinum/calomel electrode (optional) is fitted into a butyl rubber stopper. Inert gas flows continuously in through a syringe needle, and out through a stub of needle tubing, wide enough to make additions with microlitre syringes.

3. Principles of oxidation–reduction reactions and potentials

A compound, or a centre, A, which undergoes oxidation–reduction reactions is described as a redox (reduction–oxidation) compound or centre. The oxidized and reduced species, A_{ox} and A_{red}, represent a *redox couple*, A_{ox}/A_{red}. The relative strength of the redox couple in oxidation or reduction is described by the *reduction* (sometimes *redox*) *potential, E*. The units of E are volts, or millivolts.

Redox reactions can only occur if another chemical species or an electrode is involved; when a compound is reduced, something else becomes oxidized. Therefore the redox potential is always expressed relative to a standard redox compound. This is defined in terms of an electrochemical half-cell, in which the electrical potential of the redox couple is compared with that of the H^+/H_2 couple in a standard hydrogen electrode. A redox couple with a more negative potential will tend to reduce one with a more positive potential.

When the redox couple is in the standard state, $[A_{ox}] = [A_{red}]$ and the potential is the midpoint potential, sometimes represented by $E_A^{0\prime}$. The prime ' indicates that the pH is 7.0, rather than the impractical standard value of 0. For most biological work pH 7 is taken as standard and the prime is dropped. E_A^0 is a useful parameter for comparing one redox compound with another, but does not reflect the actual potential of the compound in general, since the potential depends on the concentrations of A_{ox} and A_{red}, and other factors such as the pH.

The free energy change associated with the oxidation of a compound with potential E_A by another compound, E_B, is given by:

$$\Delta G = -nF\Delta E \tag{1}$$

where $\Delta E = E_A - E_B$, all expressed in volts, F is the Faraday constant (96 485 coulomb/mol), ΔG is in J/mol. The term n is the number of electrons transferred at one time by the redox couple. The effect of concentrations of A_{ox} and A_{red} on E is given by the Nernst equation:

$$E = E_A^0 + \frac{RT}{nF} \ln \frac{[A_{ox}]}{[A_{red}]} \tag{2}$$

$$= E_A^0 + \frac{59.2}{n} \log_{10} \frac{[A_{ox}]}{[A_{red}]} \text{ at } 25°C \tag{3}$$

where R is the gas constant (8.3145 J/mol/K), and T the temperature in Kelvin. The standard temperature, T, is 25°C, 298.15 K. For most biological redox species, n is expected to be one. The value of n in the Nernst equation determines the slope of the graph; $n = 2$ has a steeper slope than $n = 1$. Note

93

that $n = 2$ only applies to a redox reaction where the simultaneous transfer of both electrons is obligatory,

$$\text{e.g. } 2H^+ + 2e^- \rightleftharpoons H_2, \text{ or } NAD^+ + H^+ + 2e^- \rightleftharpoons NADH.$$

If the two electrons may be transferred either together or separately, for example to or from a protein with two iron–sulfur clusters, the value of n is one for each centre.

More complex cases occur where a redox centre can undergo more than one redox reaction, where there are multiple centres, or where the midpoint potentials are pH-dependent. Examples of these are described in Appendix 1 (at the end of this chapter).

It may be noted that a redox potential is a measurement at equilibrium; it relates to equilibrium constants (K) and not rate constants (k). Although redox potentials give an indication of a *driving force* for a reaction, they do not directly predict the *rates*.

The thermodynamics of redox reactions in membrane bioenergetic systems, which are often coupled to transmembrane potentials, are discussed by Nicholls and Ferguson (13). In some studies on membrane vesicles, it is possible to obtain quasi-equilibrium conditions while maintaining a trans-membrane potential, generated for example by ATP. This will lead to changes in the apparent midpoint potentials, as discussed in a comprehensive review by Walz (14).

4. Equilibration methods

The midpoint potential of protein A may be estimated by equilibration with a reference compound of known midpoint potential, provided that both the protein and the reference compound have spectra which are different in the oxidized and reduced states.

The simple example given here (*Protocol 2*) is soluble cytochrome c from horse heart mitochondria. The reference compound is the mediator dye, 2,6-dichloroindophenol (DCIP). It has the advantage that the reduced form does not react rapidly with oxygen; if it did, the reaction would have to be carried out in an anaerobic spectrophotometer cell, under inert gas.

The spectrum of the system is measured with all components oxidized; then the redox state of the mixture is altered, for example by adding reductant, and the concentrations measured again. This is repeated until the system is fully reduced.

Since the difference spectra of the two components overlap, the proportions of the oxidized and reduced forms of each species ($[C_{ox}],[C_{red}]$ for cytochrome c, $[D_{ox}],[D_{red}]$ for the mediator dye) are determined by two simultaneous equations. Reduced cytochrome c has a sharp absorption band at 550 nm, while the oxidized protein has a smaller, broad absorption in this region. The proportion of reduced cytochrome may be determined in this case from the difference in absorption between 550 and 570 nm. DCIP has an absorption band around 600 nm in the oxidized state, and no absorption in this region in the reduced state. The absorption spectra of the two components may be measured with a scanning spectrophotometer, or, more conveniently, a diode array spectrophotometer. For membrane bound systems, the effects of light scattering may be compensated by the use of a dual wavelength spectrophotometer. Then:

$$E = E_C + \frac{RT}{nF} \ln \frac{[C_{ox}]}{[C_{red}]} = E_D + \frac{RT}{nF} \ln \frac{[D_{ox}]}{[D_{red}]}. \quad [4]$$

Protocol 2. Estimation of the redox potential of a cytochrome by optical absorption spectroscopy

1. Put into a spectrophotometer cuvette:
   - 1 ml cytochrome c, 2 mg/ml
   - 25 μl 1 mM potassium ferricyanide (to oxidize any reduced cytochrome c)
   - 2 ml 50 mM citrate buffer pH 6.5
   - 0.1 ml 1 mM DCIP

2. Mix well, and measure the absorbance at 550, 570, and 600 nm. These readings correspond to fully oxidized cytochrome c and DCIP. Record the difference in absorbance between 500 and 570 nm, which we will call $A(C_{ox})$, and the absorbance at 600 nm, which we will call $A(D_{ox})$.

3. Add 3 μl 10 mM ascorbate with a microlitre syringe, and mix. Allow 1 min for equilibration, then read the absorbance values $A(C)$ and $A(D)$ as before.

4. Continue additions of ascorbate and absorbance readings, until A_{600} is low and the readings are constant. Finally, add a few small crystals of sodium dithionite and take a final reading, corresponding to the fully reduced states $A(C_{red})$ and $A(D_{red})$.

5. The ratios of oxidized and reduced forms may be calculated from:

   $$\frac{[C_{ox}]}{[C_{red}]} = \frac{A(C_{red}) - A(C)}{A(C) - A(C_{ox})} \quad [5]$$

   (remembering that the sign of factors such as $A(C_{ox})$ may be negative)

   $$\frac{[D_{ox}]}{[D_{red}]} = \frac{A(D) - A(D_{red})}{A(D_{ox}) - A(D)} \quad [6]$$

6. Plot a graph of $\log_{10}([D_{ox}]/[D_{red}])$ against $\log_{10}([C_{ox}]/[C_{red}])$. The graph is a straight line, with slope n_D/n_C, and intercept on the horizontal axis $n_C/59.2 \, (E_D - E_C)$. Given that the midpoint potential of the dye is 237 mV at pH 6.5, the midpoint potential of cytochrome c may be calculated.

---

5. Spectroelectrochemical methods

The redox titration technique, and the choice of mediator titrants, is summarized by Dutton (15). The solution or membrane suspension to be investigated is kept in a stirred vessel at 25 °C with electrodes to measure the redox potential.

6. Redox cells

A number of different types of cells have been described (11, 16–18). The basic requirements are: an oxygen impermeable vessel, with provision for flushing with an inert gas; electrodes to measure the redox potential; a stirrer; provision to introduce oxidants and reductants. For techniques such as EPR spectroscopy, samples must be extracted into sample tubes without exposure to oxygen. The simple vessel described by Dutton (15) uses the pressure of nitrogen gas to propel the sample into a gas-flushed sample tube; a drawback is that back pressure can interfere with the operation of the reference electrode. Other designs allow all or part of the vessel to be evacuated.

Figure 3 shows a simple apparatus for mediator titrations for EPR spectroscopy. The electrodes are made from a piece of fairly stiff platinum foil, 5 mm square, and platinum wire. The reference is a calomel electrode (Radiometer). Sometimes it is also advantageous to incorporate a pH electrode, to check for changes in pH of the solution. The same calomel reference can be used for both, though it may be necessary to make up special plug connectors. The meter is a high impedance voltmeter; many types of pH meter have a millivolt scale for this.

The apparatus should be tested for leakage of oxygen from the atmosphere into the sample. A check for this is to use 50 μM benzyl viologen and 50 μM phenosafranin in buffer, pH 7.0, reduced with small additions of 10 mM dithionite solution until the solution just turns blue. On oxidation, the colour changes from blue to pink at a redox potential of −260 mV.

The vessel is flushed with inert gas through three long syringe needles, which are attached to a glass manifold, and this outward flow minimizes inward oxygen leakage. Oxidant and reductant are added to the vessel with screw-driven syringes, about 1 ml volume, which are capable of making additions of 1 μl or less. These are connected to fine syringe needles, such as 25 gauge (0.5 mm) spinal needles. The sample is transferred into the EPR tube with a syringe, via a small chamber which is in the inert gas-filled space, so that the needle tip is not exposed to the air during transfer. We use 250 μl gas-tight syringes with fixed needles, 0.8 mm diameter, 25 cm long; these may be ordered from Hamilton. Syringes with removable needles tend to leak oxygen and are not satisfactory.

Figure 2 shows an anaerobic cuvette adapted for titrations in a spectrophotometer. A conventional spectrophotometer may be used, if a magnetic

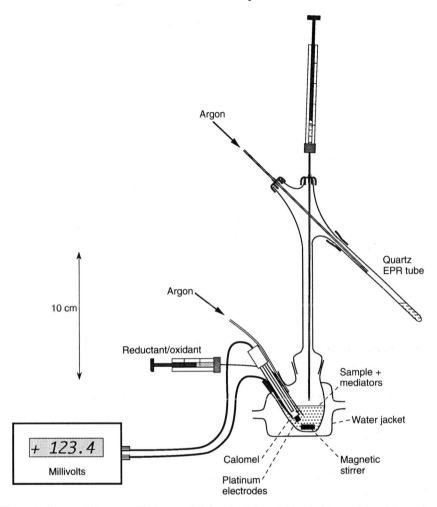

Figure 3. A vessel for controlled-potential titration of protein solutions, adapted for withdrawal of samples into EPR tubes. Inert gas flows continuously in through two syringe needles, one of which is used to purge the EPR sample tube, and out through a stub of needle tubing, wide enough to make additions with microlitre syringes, piercing a second septum.

stirrer can be fitted into the sample compartment. This design uses a combined electrode with platinum and calomel reference (e.g. Radiometer or Russell). The vessel is flushed continuously with water-saturated inert gas. Titrant solutions are added with a syringe through a small hole, which also acts as the outlet for inert gas.

In order to compensate for the optical absorption due to mediators, a diode array spectrophotometer may be used, provided that the solution does not cause significant light scattering. The spectra of the mediators may be obtained in a parallel titration and subtracted digitally.

6.1 Preparation of electrodes

6.1.1 Measuring electrodes

Mercury electrodes are not normally used because of the adsorption of proteins to the surface. Carbon electrodes may be used, such as glassy carbon (3) or graphite (4). They need to be carefully polished before use, otherwise the response is sluggish.

Platinum is the most commonly used, because of its rigidity and chemical inertness, and rapid response. Platinum wire can be fused into glass, so that it may be sealed into the wall or base of the vessel. We have constructed a titration vessel with a very small minimum volume by this technique (18). Gold is better for extremely low redox potentials, since it has a lower hydrogen overpotential than platinum; however at highly positive redox potentials it releases gold salts which may affect proteins.

Measuring electrodes tend to become coated with protein and/or lipid during use. Some hydrophobic mediators, such as naphthoquinones, which have to be dissolved in organic solvents such as ethanol, will also build up on the electrode surface. This will make equilibration slow or impossible, and necessitate thorough cleaning of the electrodes afterwards. It is better to use hydrophilic mediators such as sulfonates or carboxylates. To clean the metal electrode, it is best to remove it, and treat it with detergent, or in extreme cases strong oxidizing acids such as chromic. This is more difficult with combined electrodes, if cleaning liquids can contaminate the porous glass junction of the reference.

6.1.2 Reference electrodes

The standard hydrogen electrode is difficult to set up in practice, so secondary reference electrodes are used, such as calomel (mercury chloride/mercury) ($E = 244$ mV) or silver chloride/silver ($E \approx 200$ mV). Both types are readily available commercially. When set up under proper conditions, these have a known potential relative to the standard hydrogen electrode. In practice the potential of these references may change somewhat due to changes in the concentration of the KCl solution, so for accurate work it is necessary to calibrate them using a standard, such as quinhydrone in phosphate buffer, pH 7.0. Quinhydrone is a crystalline equimolar mixture of benzoquinone and benzohydroquinone and is added in excess to provide a saturated solution. It is worthwhile checking the potential before and after the redox titration.

6.2 Mediators

A list of useful mediators is given in *Table 1*. Most of these can be obtained from Aldrich, Fluka, or Merck. Some of them are discussed in detail by

Prince *et al.* (19). To obtain the right combination of mediators often needs some experimentation. The requirements for an efficient mediator are:

(a) A midpoint potential near (within 100 mV at most) to that of the species under investigation, at the pH of the titration.

(b) It should interact efficiently with the protein under investigation. Electron-transfer proteins have preferences for mediators, just as enzymes are specific for substrates.

(c) It should not bind to the protein and change its potential. Tight binding also means that if the concentration of mediator is less than the concentration of binding sites, the mediator will be unavailable to communicate with the electrode.

(d) In optical titrations, their optical absorption spectra should not obscure that of the protein. In some cases, the contribution of the mediators is determined by a separate titration, and the protein spectra are obtained by subtraction.

(e) For EPR titrations, it should not give interfering signals. An example is viologen dyes which give radical signals at around $g = 2.0$ in the reduced state. These must be used sparingly, and their contributions allowed for, in titrations of flavins or quinones.

Several preliminary titrations may be necessary to assess the types and amounts of mediators to use with a given system. If the redox potentials to be measured in a system span a wide-range of redox potentials, so that a large number of mediators is necessary, it may be preferable to carry out the titration in a number of stages of, say, 300 mV, using different combinations of mediators.

6.3 Achievement of redox equilibrium

It is advisable to carry out titrations in both the oxidizing and reducing directions. Poor equilibration, and too rapid changes in potential, will cause distortions of the plots of signal amplitude versus redox potential, and hysteresis so that the plots in the oxidizing and reducing directions do not coincide. If possible, plot a graph of spectral amplitude against potential, as the titration is taking place. If the apparent *n* value deviates from an expected value of one (see *Figure 4*), this may be due to co-operative effects, or to a two-electron redox process. However in this case it is wise to check for poor equilibration. This can be done by repeating the titration more slowly, by using higher concentrations of mediators, or by using other mediators.

When low-temperature spectroscopy is used to measure the redox state, it is generally assumed that the electron distribution in the frozen state is the same as that in solution. The equilibration is assumed to take place at 25°C. However it has been found that in some complex multicentred proteins such

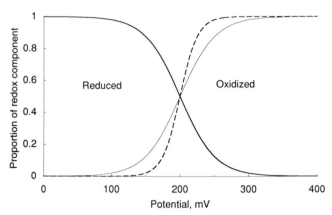

Figure 4. Theoretical redox curves of the redox states of a centre, showing the proportions of oxidized and reduced species, for a redox couple with $n = 1$, and oxidized species, with $n = 2$ (dotted).

as xanthine oxidase, some electron redistribution can take place during cooling to cryogenic temperatures. Thus the potentials estimated at low temperatures deviate from those estimated at room temperature (20).

The method of adjusting the potential of a protein solution is similar for both optical and EPR measurements (*Protocol 3*). For EPR spectroscopy, samples have to be withdrawn and frozen. The method is suitable for iron–sulfur proteins, flavoproteins, or other proteins where the redox state is readily observed by EPR spectroscopy. It is assumed that facilities are available to do this; the methods have been described elsewhere (e.g. 10, 18, 21).

For quantitative work the internal diameter of the quartz EPR sample tubes must be known since the volume measured is proportional to the square of the diameter. Precision bore quartz tubes are available from the Willmad Glass Co. Alternatively cheaper tubes can be calibrated, and the EPR signals of different samples corrected for their internal diameter. Calibration can be done to an accuracy of $\pm$ 2% by filling the tubes with water to a known length, and weighing the water.

Protocol 3. Estimation of the redox potential of a protein by EPR spectroscopy

Equipment and reagents

- Redox cell (see *Figure 3*) and gassing manifold (see *Figure 1*)
- Quartz EPR tubes (about 20 per titration), labelled with sample number and tube diameter
- Hamilton syringe, gas-tight, 250 µl, with fixed 15 cm needle (about 21 gauge, 0.8 mm outer diameter)
- Digital pH/millivolt meter
- Solid quinhydrone (e.g. Aldrich)
- Redox mediator solutions, 10 mM in water (see *Table 1*)
- Fresh sodium dithionite solution, 100 mM in 100 mM Tris–HCl pH 9
- Potassium ferricyanide solution, 200 mM

5: Redox states and potentials

Method

1. To calibrate the electrode potential, put about 3 ml standard phosphate buffer, pH 7.0, in the vessel and add about 0.2 g solid quinhydrone. Stir the mixture well (the solid does not have to dissolve completely), and read the potential. The standard potential of quinhydrone at pH 7.0 and 25°C is +286 mV. Hence if, for example, the meter reads +42 mV, the reference electrode potential is +244 mV, and this value must be added to all readings.

2. If the protein has an enzymic activity, it is advisable to measure this both before and after the titration, if the mediators do not interfere.

3. Set the temperature of the vessel to the desired temperature (the standard is 25°C).

4. Put the solution of the protein in the redox vessel, and bring to the required volume with buffer. For the redox medium, choose a zwitterionic buffer which does not change its pH on freezing (22).

5. Add mediator solutions. The concentration should be found by trial and error, but is typically 20–50 µM.

6. Flush the solution with gas while stirring, for 5–10 min.

7. Put an EPR tube on the side arm and flush with inert gas, as indicated in *Figure 3*.

8. Take a reading of the redox potential of the solution.

9. Empty and fill the gas-tight syringe with the mixture of protein and mediators from the vessel. Repeat ten times, or until the reading on the meter is constant (be careful not to touch the electrode with the syringe needle). Transfer a 150 µl sample into the EPR tube, and freeze.

10. Remove the EPR tube and store in liquid nitrogen for EPR measurements. Replace the EPR tube with a new one and begin to flush with gas.

11. Adjust the potential of the solution by addition of small aliquots of oxidant or reductant, until the potential is as required. Wait the appropriate equilibration time (typically 5 min), and remove a sample as above.

12. For a reasonable titration plot, points should be taken at intervals of 20–30 mV around the midpoint.

13. For most precise measurements, make a note of the volume of sample withdrawn, and the volumes of titrant added, so correction may be made for dilution of the sample. If the material is likely to deteriorate with time, the time of withdrawing samples should also be recorded. It is possible to make an approximate allowance for the deterioration by taking samples with maximal signal amplitude at the beginning and end of the titration, and assuming a linear rate of decay.

The potential is changed by addition of small volumes of solutions of oxidizing and reducing agents. The solutions in the syringes should have a similar density to the main contents of the vessel, otherwise titrant solutions will tend to 'fall out of' the syringe needles and mix with the bulk solution.

The spectrum of each sample must be measured under standard conditions of temperature, sample size, tube diameter, and position in the cavity. There is a considerable advantage in recording the EPR spectra at the same time as the titration is taking place. This makes it much easier to narrow down the range of potential to be investigated, and to check for deviations from expected redox behaviour caused, for example, by inadequate equilibration.

7. Fitting of data to estimate redox potentials

After measuring the spectral amplitude as a function of redox potential, the results may be fitted to theoretical curves of redox state versus potential. The general types of redox process, and the expected plots, may be calculated from the Nernst equation, as shown in Appendix 1.

For a single redox species, the variables to be fitted are the amplitudes of the spectrum in the fully oxidized and fully reduced states and the midpoint potential, E^0. The theoretical curves may be calculated, and plotted, with the data points, on a computer, using a general calculation/plotting program such as a spreadsheet. The variables are adjusted to obtain the best fit as judged by eye. Alternatively, it is possible to use a non-linear least-squares fitting routine. If so, it is important to check that all the data points are good. A 'wild' point, caused for example by accidental exposure of a sample to oxygen, can readily be discarded by inspection, otherwise it may severely affect the statistical analysis.

An advantage of EPR spectroscopy is that the integrated intensity of the spectrum is proportional to the number of unpaired electrons. Quantitation in EPR may be achieved by double integration of the signal, measured under non-saturating conditions, and comparing with a standard sample of a paramagnet such as a Cu(II) solution (23). This is particularly true for $S = \frac{1}{2}$ systems, though systems with $S > \frac{1}{2}$ may also be treated quantitatively (18). In favourable circumstances the total integrated intensity may be followed as a function of redox potential, and thus the relative numbers of electrons in each redox centre may be estimated.

For electron transfer proteins or membrane bioenergetic systems, which have multiple redox centres, complex and unexpected behaviour may occur. For example, if there are two redox centres A and B, reduction of the oxidized form [AB] may cause a spectral feature due to the species [A⁻B] to appear. On further reduction, this may shift or disappear, because the spectra of [AB⁻] and [A⁻B⁻] are different. In such cases it is necessary to set up a model for the redox centres and their interactions. The formalism used in Appendix 1 may be elaborated to set up the equations, which must take into account all possible combinations of redox states of the centres, and the electron transfer steps connecting them. An example is the redox states of xanthine oxidase, which contains two iron–sulfur clusters, flavin and molybdenum, giving a total of 36 possible intermediates (24). The curves used to fit the experimental data may be a combination of effects, for example a spectrum due to interaction between the intermediate Mo(V) state and a reduced iron–sulfur cluster.

Acknowledgements

I thank my numerous colleagues over the years, with whose assistance these techniques were developed. The work was supported by grants from the Science and Engineering Research Council.

References

1. Wilson, G. S. (1978). In *Methods in enzymology* (ed. S. Fleischer and L. Packer), Vol. 54, pp. 396–410. Academic Press, New York.
2. Chamorovsky, S. K. and Cammack, R. (1982). *Photobiochem. Photobiophys.*, **4**, 195.
3. Hagen, W. R. (1989). *Eur. J. Biochem.*, **182**, 523.
4. Armstrong, F. A. (1990). *Struc. Bonding*, **72**, 137.
5. Sucheta, A., Ackrell, B. A. C., Cochran, B., and Armstrong, F. A. (1992). *Nature*, **356**, 361.
6. Kay, C. J., Barber, M. J., and Solomonson, L. P. (1988). *Biochemistry*, **27**, 6142.
7. Sweetser, P. B. (1967). *Anal. Chem.*, **39**, 979.
8. McKenna, C. E., Gutheil, W. G., and Song, W. (1991). *Biochim. Biophys. Acta*, **1075**, 109.
9. Thomson, A. J., Robinson, A. E., Johnson, M. K., Cammack, R., Rao, K. K., and Hall, D. O. (1981). *Biochim. Biophys. Acta*, **637**, 423.
10. Beinert, H., Orme-Johnson, W. H., and Palmer, G. (1978). In *Methods in enzymology* (ed. S. Fleischer and L. Packer), Vol. 54, pp. 111–32. Academic Press, New York.
11. Paulsen, K. E., Stankovitch, M. T., and Orville, A. M. (1993). In *Methods in enzymology* (ed. J. L. Riordan and B. L. Vallee), Vol. 227, pp. 396–411. Academic Press, New York.
12. Dixon, M. (1971). *Biochim. Biophys. Acta.*, **226**, 241.
13. Nicholls, D. G. and Ferguson, S. J. (1992). *Bioenergetics 2*, pp. 48–52. Academic Press, London.
14. Walz, D. (1979). *Biochim. Biophys. Acta*, **505**, 279.
15. Dutton, P. L. (1978). In *Methods in enzymology* (ed. S. Fleischer and L. Packer), Vol. 54, pp. 411–34. Academic Press, New York.
16. Harder, S. R., Feinberg, B. A., and Ragsdale, S. W. (1989). *Anal. Biochem.*, **181**, 283.
17. Kay, C. J. and Barber, M. J. (1990). *Anal. Biochem.*, **184**, 11.

18. Cammack, R. and Cooper, C. E. (1993). In *Methods in enzymology* (ed. J. L. Riordan and B. L. Vallee), Vol. 227, pp. 353–84. Academic Press, New York.
19. Prince, R. C., Linkletter, S. J. G., and Dutton, P. L. (1981). *Biochim. Biophys. Acta*, **351**, 132.
20. Porras, A. G. and Palmer, G. (1982). *J. Biol. Chem.*, **257**, 11617.
21. Cammack, R. (1993). In *Methods in molecular biology*, Vol. 17: Spectroscopic methods and analyses (ed. C. Jones, B. Mulloy, and A. Thomas), pp. 327–44. Humana Press, Totowa, NJ.
22. Williams-Smith, D. L., Bray, R. C., Barber, M. J., Tsopanakis, A. D., and Vincent, S. P. (1977). *Biochem. J.*, **167**, 593.
23. Randolph, M. L. (1972). In *Biological applications of electron spin resonance* (ed. H. M. Swartz, J. R. Bolton, and D. C. Borg), pp. 119–53. Wiley-Interscience, New York.
24. Olson, J. S., Ballou, D. P., Palmer, G., and Massey, V. (1974). *J. Biol. Chem.*, **249**, 4363.

Appendix 1
Equations for different types of oxidation–reduction reactions

In order to determine the midpoint potential, E^0, of a species, the proportion of its oxidized or reduced form, determined spectroscopically, is plotted against the applied potential, E. From the Nernst equation it is possible to derive the theoretical curves of concentration of redox species against applied potential, for various cases.

Case 1. Simple redox centre

RS 1

$$\text{Proportion of } [A] = \frac{x}{(x + e_A)} \quad [A^-] = \frac{e_A}{(x + e_A)} \tag{7}$$

$$\text{where } x = \exp\left(\frac{nF}{RT} \cdot E\right) e_A = \exp\left(\frac{nF}{RT} \cdot E_A^0\right).$$

A plot of the concentration of redox species versus redox potential is an S-shaped curve (see *Figure 4*). Alternatively, a straight line graph can be obtained by plotting log ([A]/[A⁻]) against E (see *Figure 5*).This crosses the axis at $E = E_A$. When fitting the line to the data, the points near to E_A are expected to have less scatter than those at the ends of the line.

Figure 6 shows the calculated redox curve for the superimposition of four redox species. Note that it is difficult to distinguish the two species with potentials separated by only 50 mV.

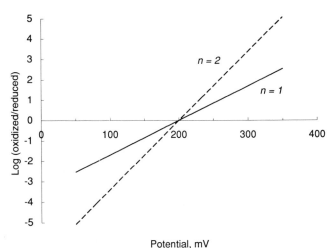

Figure 5. Semi-log plot of log $[A_{ox}]/[A_{red}]$ versus potential.

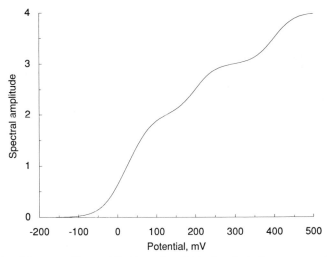

Figure 6. Plot of the sum of four oxidized species which give rise to the same spectral feature. Calculated for midpoint potentials 0, 50, 200, 400 mV, all with equal amplitudes, and $n = 1$.

Case 2. pH effects

If the reduction of a centre by an electron is accompanied by binding of a proton (not necessarily to the same site as the electron), i.e. it is effectively reduced by hydrogen, the potential is pH-dependent, changing by 59.2 mV per pH unit, since:

$$E = E^0 + \frac{RT}{nF} \ln \frac{[A]}{[A^-]} + \frac{2.303RT}{nF} \cdot \text{pH}. \qquad [8]$$

105

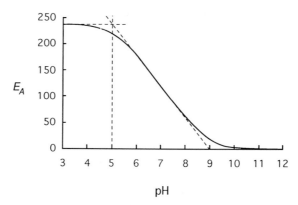

Figure 7. Example of the effect of pH on the midpoint potential, where the pK_a of a redox centre changes from 5 to 9 on reduction.

Another case is where reduction of a centre alters the pK_a of a group on a protein, so that a change in protonation of the protein accompanies the redox reaction.

$$
\begin{array}{ccc}
& E_A^0 & \\
\text{A} & \rightleftharpoons & \text{A}^- \\
pK \uparrow\downarrow & & \uparrow\downarrow pK' \\
& E_A'^0 & \\
\text{AH}^+ & \rightleftharpoons & \text{AH}
\end{array}
$$

In this case, the effect of pH on E^0 depends on the value of pK_a relative to the pH (see *Figure 7*). Note that in the region between the two pK_a values, the potential changes with $E^0/\text{pH} \approx 59.2$ mV.

This is a special case of the more general process, where the binding constant of a ligand, L, is dependent on the redox state. Reciprocally, the binding of L will affect the midpoint potential.

Case 3. A single redox centre which undergoes two successive redox reactions

Examples are flavins and quinones, which undergo reduction via a semi-quinone state, or molybdenum: $\text{Mo(VI)} \longleftrightarrow \text{Mo(V)} \longleftrightarrow \text{Mo(IV)}$.

The general redox reaction is of the form:

$$
\begin{array}{ccccc}
& E_1^0 & & E_2^0 & \\
\text{A} & \rightleftharpoons & \text{A}^- & \rightleftharpoons & \text{A}^{2-}
\end{array}
$$

where A, A^- and A^{2-} represent the oxidized, semi-reduced, and fully reduced forms respectively.

If we consider measurements by EPR spectroscopy of flavins, quinones, or molybdenum, the semi-reduced, intermediate species A^- is the only species detectable.

$$\text{Proportion of } [A] = \frac{x^2}{x^2 + x.e_1 + e_1.e_2} \qquad [9]$$

$$[A^-] = \frac{x.e_1}{x^2 + x.e_1 + e_1.e_2} \qquad [10]$$

$$[A^{2-}] = \frac{e_1.e_2}{x^2 + x.e_1 + e_1.e_2} \qquad [11]$$

$$\text{where } x = \exp\left(\frac{nF}{RT}.E\right) \qquad e_1 = \exp\left(\frac{nF}{RT}.E_1^0\right)$$

$$e_2 = \exp\left(\frac{nF}{RT}.E_2^0\right)$$

The plot of signal versus E for A^-, the EPR-detectable species, is a bell-shaped curve (see *Figure 8*). The peak occurs at a potential midway between E_1^0 and E_2^0. The amplitude of the signal at the peak depends on the difference between E_1^0 and E_2^0. If $E_1^0 = E_2^0$, the three curves for the concentrations of A, A^- and A^{2-} meet at $(E_1^0 + E_2^0)/2$, each with amplitude one-third of the maximum. If $E_1^0 > E_2^0$, the intermediate form has a maximum intensity greater than this. If $E_1^0 < E_2^0$, it is less (see *Figure 8*). The maximum intensity of the intermediate species A^- always occurs at $[E_1^0 + E_2^0)/2]$. In order to obtain the separate values, E_1^0 and E_2^0, it is necessary to make a quantitative determination of the intermediate species, as a proportion of total $[A]$. The difference $(E_1^0 - E_2^0)$ may be estimated by using Equation 3.

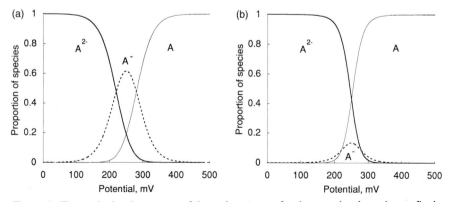

Figure 8. Theoretical redox curves of the redox states of redox species A, such as a flavin, that has two sequential redox reactions. Calculated for (a) $E_1^0 = 250$ mV, $E_2^0 = 200$ mV, (b) $E_1^0 = 250$ mV, $E_2^0 = 200$ mV.

Case 4. Two redox centres, A and B

For example two haems or iron–sulfur clusters in a protein.

$$
\begin{array}{ccc}
& E_A^0 & \\
AB & \rightleftharpoons & A^-B \\
E_B^0 \updownarrow & & \updownarrow E_B^{0'} \\
& E_A^{0'} & \\
AB^- & \rightleftharpoons & A^-B^-
\end{array}
$$

It may be noted that the EPR spectrum of A^- in A^-B may be different from that in A^-B^-. Thus, the spectrum of A^-B appears during reduction, then disappears as it is replaced by that of A^-B^-. The same applies to the spectrum of B^-, which may change when A is reduced. The relative amplitudes of these intermediate species depend on the relation between E_A^0 and E_B^0. **Co-operativity**. The effect of reduction of centre B may be to alter the midpoint potential of A, from E_A^0 to a $E_A^{0'}$ (here the prime ' denotes a modified potential, not a standard potential). If the effect of reduction of B is to facilitate the reduction of A (so $E_A^{0'}$ is more positive than E_A^0), this is positive co-operativity. The effect is mutual, so that reduction of centre A will make E_B^0 more positive by the same amount.

$$\text{Proportion of AB} = \frac{x^2}{x^2 + x.e_A + x.e_B + c.e_A.e_B} \tag{12}$$

$$A^-B = \frac{x.e_A}{x^2 + x.e_A + x.e_B + c.e_A.e_B} \tag{13}$$

$$AB^- = \frac{x.e_B}{x^2 + x.e_A + x.e_B + c.e_A.e_B} \tag{14}$$

$$A^-B^- = \frac{c.e_A.e_B}{x^2 + x.e_A + x.e_B + c.e_A.e_B} \tag{15}$$

where $x = \exp\left(\dfrac{nF}{RT}.E\right)$ $e_A = \exp\left(\dfrac{nF}{RT}.E_A^0\right)$

$$e_B = \exp\left(\frac{nF}{RT}.E_B^0\right)$$

$$c = \exp\left(\frac{nF}{RT}.(E_A^{0'} - E_A^0)\right) = \exp\left(\frac{nF}{RT}.(E_B^{0'} - E_B^0)\right).$$

The two intermediate species, A^-B and AB^-, both show bell-shaped curves with maxima at $(E_1^0 + E_2^0)/2$. An example is shown in *Figure 9*. If $E_A^0 = E_B^0$, and there is no co-operativity, the four lines meet at E_A^0, each

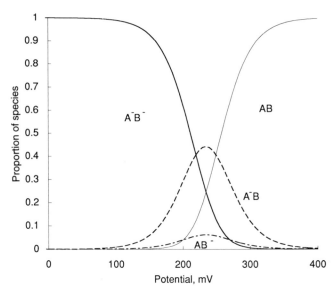

Figure 9. Plots for two redox species, A and B, which interact. Calculated for $E_A = 180$ mV; $E_B = 100$ mV; co-operativity = 50 mV.

with one-quarter of the maximum intensity. The effect of *positive* co-operativity is to increase the slope of the curves, so that the apparent value of n is more than one. In the extreme case, where reduction of one centre inevitably leads to the reduction of the other, the effective n value is two. The effect of *negative* co-operativity, so that reduction of A makes E_B more negative, and vice versa, is to make the slope of the curve less steep. If one is observing, for example A^- (i.e. the sum of A^-B and A^-B^-), the effect may be to observe two apparent waves, which can resemble two redox components.

6

Control and regulation in bioenergetics

GUY C. BROWN

1. Introduction

Control and regulation have traditionally been studied in a non-quantitative way, in order to determine where and how metabolic rates and levels are changed. The analysis of 'control' is concerned with *where* within a system the rates of pathways or levels of intermediates can be changed, i.e. which enzymes or transporters limit the system rates or levels. The analysis of 'regulation' is concerned with *how* a physiological change in rate or level is brought about. A variety of quantitative approaches are now available to help estimate the extent to which different metabolic components control rates and levels, and the extents to which different mechanisms are involved in producing a response to some effector or physiological change. These quantitative approaches have in turn changed our concepts of control and regulation, and how we should talk about them. Bioenergetics has had a traditional emphasis on quantitative analysis, and thus these approaches have rapidly been applied to a variety of bioenergetic systems, including bacteria, mitochondria, chloroplasts, and proteoliposomes.

This chapter outlines how to measure control and regulation in bioenergetic systems using approaches designed to answer the following types of question:

(a) Which enzymes limit the rate of a pathway, and to what extent?

(b) Which component steps or rate constants within a single isolated enzyme limit the rate, and to what extent?

(c) Which mechanisms are involved in producing the response of a pathway to some effector or change, and to what extent are different mechanisms involved?

(d) How does a particular membrane process depend on $\Delta\psi$ and ΔpH?

'Enzyme' is used in this chapter as shorthand for enzymes, transporters, ATPases, or any protein with a rate. As several of the approaches used in this chapter depend on an understanding of metabolic control theory, we first outline this theory.

2. Control analysis

2.1 Metabolic control theory

Metabolic control theory, also known as 'metabolic control analysis', has been reviewed together with its experimental application by Fell (1). *Table 1* lists the definitions, theorems, and useful relations for control over flux. There are equivalent definitions and different theorems for control over concentrations or levels (1). The extent to which an enzyme limits the steady state rate of a pathway is quantified as a 'flux control coefficient'. This has two definitions:

(a) A simple definition in terms of the sensitivity of the pathway flux to the enzyme's concentration.

(b) A general definition in terms of the ratio of the sensitivity of the pathway flux and the 'isolated' enzyme's rate to a specific effector of that enzyme.

Normally these two definitions are equivalent. The response coefficient is a measure of the sensitivity of the pathway flux to an external effector. The elasticity coefficient (or 'elasticity' for short) is a measure of the sensitivity of the 'isolated' enzyme to an effector. The effector might be stimulator, inhibitor, substrate, product, or physical change. 'Internal effectors' are effectors within the system (such as metabolic intermediates); 'external effectors' are effectors not included in the system (such as inhibitors).

The control, response, and elasticity coefficients are related via the various definitions and theorems of *Table 1*. The summation theorem states that sum of the control coefficients of all the enzymes in the system is equal to one. The connectivity theorem implies that if two enzymes in a linear pathway are connected via a common metabolite (or interdependent metabolites), then the relative control exerted by those enzymes is determined by their relative sensitivities (elasticities) to their common intermediate. This not only applies to two adjoining enzymes, but also to the two adjoining ends of the pathway, or to the control exerted by a number of different pathways connected by a common intermediate. Although in the latter cases the ('overall') control coefficients refer to the control exerted by groups of enzymes, and the ('overall') elasticity coefficients refer to the sensitivities of these groups of enzymes to an effector. Similarly the branching theorem relates the control exerted by all the enzymes in a branch of a pathway to the relative flux of the branch.

It is important to bear in mind a number of limitations of metabolic control theory:

(a) The theory deals with steady states only.

(b) The theory deals with infinitesimal changes only, and thus the coefficients can not describe or predict the effects of large changes.

6: Control and regulation in bioenergetics

Table 1. Metabolic control theory for fluxes

Definitions

[1] Flux control coefficient [a,b] (simple definition) $C_i^j = \dfrac{\partial J_j}{\partial E_i} \cdot \dfrac{E_i}{J_j}$

[2] Response coefficient [a,c] $R_k^i = \dfrac{\partial J_j}{\partial M_k} \cdot \dfrac{M_k}{J_j}$

[3] Elasticity coefficient [a,d] $\varepsilon_k^i = \dfrac{\partial v_i}{\partial M_k} \cdot \dfrac{M_k}{v_i}$

[4] Flux control coefficient [a,c,d] (general definition) $C_i^j = R_k^i/\varepsilon_k^i = \dfrac{\partial J_j}{\partial M_k \cdot J_j} \Big/ \dfrac{\partial v_i}{\partial M_k \cdot v_i}$

Theorems

[5] Summation theorem [e] $\displaystyle\sum_{i=1}^{n} C_i^j = 1$

[6] Connectivity theorem [e,f] $\displaystyle\sum_{i=1}^{n} C_i^j \cdot \varepsilon_x^i = 0$

[7] Branching theorem [g] $C_k^j/C_l^j = J_k/J_l$

Relations

[8] Combined response [e,h] $R_k^i = \displaystyle\sum_{i=1}^{n} C_i^j \cdot \varepsilon_k^i$

[9] Control by branches [i] $C_i^j = \dfrac{J_i \cdot \varepsilon_x^i}{-\displaystyle\sum_{a=1}^{n} (J_a \cdot \varepsilon_x^a)} + \delta_{ij}$

[a] J_j is the steady state rate (flux) of pathway j.
[b] E_i is the concentration of enzyme i.
[c] M_k is the concentration of effector k.
[d] v_i is the steady state rate of enzyme i when 'isolated'. 'Isolated' means kinetically isolated from the pathway by holding substrate, product, and other effector concentrations constant. The general definition of the flux control coefficient holds only when the effector specifically acts on enzyme i, and the elasticity coefficient is for the conditions of the pathway (i.e. same concentrations of substrates, products, other effectors, temperature, etc.). The effector must be an external effector, not a metabolite internal to the system.
[e] n is the total number of enzymes in the system.
[f] The elasticities are for an internal effector x, i.e. an intermediate within the system.
[g] J_j, J_k, and J_l are fluxes of three branches, connected only at a single intermediate. C_k is the sum of the control coefficients of all the enzymes of branch k, over the flux of branch j.
[h] The elasticities are to an external effector k measured on the 'isolated' enzyme with the same metabolite concentrations and conditions as for the response coefficient.
[i] This is the control exerted by all the enzymes of branch i over the flux of branch j. The branches interact only at their common intermediate x. Fluxes are positive towards the intermediate, and negative away from the intermediate. The sum is over all branches, δ_{ij} is a constant equal to one when i is j (i.e. when looking at control by branch over its own flux), and otherwise equal to zero.

(c) The coefficients refer to a single condition only, and will be different in different conditions.

(d) The theory (in its basic form) assumes freely-diffusible intermediates between the enzymes.

2.2 Experimental design

Before measuring control coefficients in a pathway certain decisions need to be made.

2.2.1 Defining the system or pathway

The system under study is operationally defined by which metabolites are held constant experimentally. These metabolites are the boundary metabolites, and are often chosen to be the substrates and products of the pathway. The system under study will be changed by changing the boundary metabolites. Thus it is necessary to consider what system you want to know the control distribution for, and which metabolites it is practically possible to hold constant. Theoretically the concentrations of boundary metabolites should be held constant, but since there is a flux from or to these metabolites this is often impossible, unless using a perfusion system. In practice what is often done is to add large amounts of the substrates and products of the pathway, so that the change in concentrations of the substrates and products during the time course of the experiment is negligible compared to the elasticity coefficients of the component enzymes to these metabolites.

An alternative is to add to the system an excess of an enzyme (or enzymes) and metabolites to interconvert substrates and products. For example, in order to fix the ATP/ADP ratio during ATP synthesis with isolated mitochondria creatine kinase plus creatine and creatine phosphate can be added to the system. A large excess of creatine kinase is required to ensure that the control coefficient of creatine kinase over ATP synthesis is zero. Large amounts of creatine and creatine phosphate are also required to ensure that their ratio does not change significantly during the experiment. Different added creatine/creatine phosphate ratios can be used to set different ATP/ADP ratios (2).

2.2.2 Choosing boundary conditions

The distribution of control coefficients between the different enzymes of the system is known as the 'control distribution'. The control distribution depends on the conditions used: the concentrations of substrates, products, and other effectors, as well as temperature, ionic strength, etc. These conditions must be chosen such that the control distribution is measured in conditions relevant to the questions being asked. Often the most relevant conditions are the physiological conditions. It is a common mistake to choose unphysiological conditions of temperature and substrate, product, and other effector concentrations. Often it is useful to measure control over a range of effector concentrations which correspond to the physiological range. For example control over mitochondrial respiration is usually measured in a range of steady states from state 4 (with a high ATP/ADP ratio) to state 3 (with a low ATP/ADP ratio).

2.2.3 Choosing the time domain

When conditions are changed different processes may reach a new steady state at very different rates. For example enzyme intermediate levels within an enzyme adjust to changes in enzyme effector levels very rapidly (< 1 sec), whereas the pools of intermediate metabolites within pathways may take seconds or minutes to reach a new steady state, and there are often processes that relax more slowly (minutes to hours). These slower processes include synthesis and breakdown of cofactors (e.g. NAD^+ and NADH) and moiety-conserved cycle metabolites (e.g. citric acid cycle intermediates), slow transport processes (e.g. slow volume changes in cells or organelles), and changes in gene expression. Thus complex metabolic systems rarely, if ever, reach a true steady state. Rather there are different time domains with different pseudo-steady states. Thus the control coefficients are a function of time. Usually we are interested in the control coefficients over a particular time domain (often second to minutes), but we need to be aware of the extent to which this is a pseudo-steady state, and that methods of estimating control coefficients over different time-scales may give different results. For example the control coefficient of cytochrome oxidase over mitochondrial respiration is likely to be different when the effects of an acute change in its activity (e.g. due to cyanide inhibition) are compared to chronic changes (e.g. due to manipulating its level by genetic methods).

2.2.4 Which control coefficients to measure and how

There are two different basic approaches to measuring control coefficients:

- the 'bottom-up' approach measures control coefficients of individual enzymes
- the 'top-down' (or 'modular') approach measures control coefficients of groups of enzymes or parts of metabolism as a unit

Multiple application of either approach results in the same information, and they may be used to complement each other. Which approach is used depends on:

- what type of question is being asked
- what experimental resources are available

If the question is 'how much control does a particular enzyme have?' then the bottom-up approach must be used; if the question is 'how is the control distributed between different parts of the system?' then the top-down approach should be used.

There are also two different methodologies for measuring control co-efficients:

(a) Direct manipulation of the enzyme (or enzymes) of interest to measure the response coefficients (and using Equations 1 or 4 of *Table 1* to estimate the control coefficients).

(b) Indirect estimation of control from measured elasticities to intermediates within the system (using Equation 6 of *Table 1*).

The two methodologies require different experimental resources. The direct method requires a method of specifically changing an enzymes concentration or a specific inhibitor of the enzyme, and some way of knowing how much it has been inhibited. The indirect method does not require very specific manipulations, but it does normally require accurate measurements of the levels of intermediates in the system, and usually some prior knowledge of what interacts with what (i.e. the structure of the system). Both methods require accurate measurements of the flux (or fluxes) within the system. The top-down approach can only use the second, indirect method, and requires prior knowledge of all the interactions between the parts of the system being studied.

2.3 Control coefficients of individual enzymes

A variety of methods have been used to estimate individual control co-efficients, i.e. the bottom-up approach (reviewed in ref. 1).

2.3.1 Genetic methods

A variety of genetic methods have been used to alter enzyme activity and thus measure control coefficients (1), including:

- breeding of homozygote and heterozygotes for different alleles of the gene
- selection or creation of mutants with altered enzyme activity
- transformation with plasmid with extra copy (or copies) of the gene, in some cases under the control of an artificial inducer
- transformation with a gene expressing antisense RNA to inhibit expression of a particular gene

Some of these methods suffer from a difficulty in producing the relatively small changes in enzyme activity required to estimate control coefficients.

2.3.2 Titration with specific inhibitors

Specific inhibitors can be used to estimate control coefficients if the response and elasticity coefficients of the enzyme to the inhibitor can be measured or estimated in the same conditions (see Equation 4 in *Table 1*). The response coefficient can be estimated from the initial slope of a graph of pathway flux against inhibitor concentration. However, estimating the initial slope by fitting a line to the initial part of the graph can be unreliable, while fitting a polynomial curve of low degree or an inhibitor equation to a larger range of the data is more reliable (3). The elasticity coefficient of the enzyme to the inhibitor can be estimated (without further experiments) if the following are known:

- the type of inhibition
- the inhibitor constants
- the local concentration of the inhibitor in the system

Equations for estimating elasticities for irreversible inhibitors and competitive and non-competitive inhibitors have been given (4, 5).

For a specific irreversible inhibitor:

$$C = - \frac{\mathrm{d}J}{\mathrm{d}I} \cdot \frac{I_{max}}{J^0}$$

where J^0 is the uninhibited flux, and $\mathrm{d}J/\mathrm{d}I$ is the initial slope of a graph of flux (J) against inhibitor concentration (I). This slope is negative, and a negative sign appears in the equation to turn this into a positive value. I_{max} is the concentration of inhibitor just sufficient to cause maximum inhibition. The derivation of the equation assumes that inhibition is independent of substrate and product concentration, and inhibition is proportional to inhibitor concentration. It is not possible to use this equation in complex systems where high levels of inhibition cause secondary effects on the pathway not present at low levels of inhibition. Where the inhibitor is not completely irreversible, it may be necessary to fit an appropriate equation to the whole curve (3).

For a specific non-competitive inhibitor:

$$C = - \frac{\mathrm{d}J}{\mathrm{d}I} \cdot \frac{K_I}{J^0}$$

where K_I is the K_I for the inhibitor measured on the 'isolated' enzyme in the same conditions as in the system.

For a specific competitive inhibitor:

$$C = - \frac{\mathrm{d}J}{\mathrm{d}I} \cdot \frac{K_I (1 + S/K_M)}{J^0}$$

where S is the concentration of the substrate that competes with the inhibitor and K_M is the K_M for that substrate.

However, there are several pitfalls in applying a K_I measured on an isolated enzyme to an intact system. First the K_I must be measured in exactly the same conditions. Secondly it is the free concentration of the inhibitor at the site of inhibition which is the relevant concentration, so that any binding or transport of the inhibitor causes problems in estimating the relevant concentration.

Thus competitive inhibitors and inhibitors that are actively transported should be avoided. For tight binding non-competitive inhibitors, the elasticity to the inhibitor may be estimated from the titration curve of the pathway flux in the intact system using the whole curve up to high levels of inhibition. Equations can be fitted to the whole curve to estimate both the elasticity and response coefficients (3). Potential errors in this method and ways of overcoming them are considered in ref. 3.

Protocol 1 is for measuring the control of the ANT over respiration in state 3. It can be adapted to measure control in other respiratory states either by:

- changing the hexokinase concentration (5)
- replacing hexokinase and glucose by creatine kinase and different ratios of creatine to creatine phosphate (2)

Using lower hexokinase concentrations sets different rates of ATP turnover and respiration, but hexokinase then becomes part of the system with its own control coefficient (which can be estimated using Equation 1 of *Table 1*). State 4 can be obtained in the absence of hexokinase (and with ATP replacing ADP in the medium). The creatine kinase system can be used to set particular external phosphorylation potentials and excludes control by creatine kinase because it is always present in excess.

Protocol 1. Flux control coefficient of the adenine nucleotide translocator (ANT) in isolated mitochondria[a]

Equipment and reagents

- Oxygen electrode, electrode meter, thermostatted electrode vessel, water-bath, and magnetic stirrer for vessel (see Chapter 1; Rank Brothers)[b]
- Chart recorder or computer for data acquisition
- Pipettes and tips, 10 μl Hamilton syringes
- Isolated mitochondria of known protein concentration (6)

- 100 mM KCl, 20 mM Hepes, 10 mM K_2HPO_4, 3 mM $MgCl_2$, 1 mM EGTA, plus 20 mM glucose, 10 mM succinate, 1 mM malate, 1 mM ADP, 5 μM rotenone, and hexokinase (about 0.2 U/ml) brought to pH 7.2 and 37°C[c]
- Carboxyatractyloside (50 μM stock)

Method

1. Put 5 ml of incubation medium in oxygen electrode vessel thermostatted at 37°C.[b]
2. Add 1 mg of mitochondrial protein/ml of incubation medium.[d]
3. Put vessel top on and exclude air bubbles.
4. Wait until oxygen electrode trace is linear (up to 5 min).[d,e]
5. Add about 50 pmol carboxyatractyloside/mg protein and wait until oxygen consumption is linear and measurable (1–2 min).[d,e]

---

6. Repeat step 5 until respiration is fully inhibited.

7. Repeat steps 1–6 sufficient times to estimate errors (about three times).

8. Measure rates of respiration from chart recorder (or computer) and plot respiration rate versus concentration of carboxyatractyloside.

9. If initial slope of graph is linear use the equation for irreversible inhibitors above to estimate control coefficient of ANT over respiration; if not linear fit low-order polynomial to first half of graph or a non-competitive inhibitor equation to whole graph (see ref. 3).

[a] Adapted from ref. 5.
[b] If you only have a small amount of mitochondria, use a small incubation vessel and 1 ml of incubation medium.
[c] This medium can obviously be adapted according to what substrate type and concentrations, ADP and ion levels, etc. you want. Glucose and hexokinase are present to induce ATP turnover at a rate dependent on how much hexokinase is added.
[d] You will have to adjust these figures according to the type and activity of your mitochondria.
[e] Check first that oxygen consumption would remain linear for the whole course of the experiment. If not change the conditions such that it is reasonably linear (e.g. lower the temperature, increase substrate concentrations), add a single aliquot of the inhibitor at a particular time point when oxygen consumption is pseudolinear, and repeat with another incubation and inhibitor concentration.

i. Inhibitors used in isolated mitochondria (6)

The inhibitors normally used to measure control coefficients in isolated mitochondria (with I_{max} values for heart mitochondria) are: carboxyatractyloside (irreversible, I_{max} 1360 pmol/mg protein) for the adenine nucleotide translocator; oligomycin (irreversible, but slow, I_{max} 216 pmol/mg protein) for the ATP synthase; mersalyl (irreversible, but slow and not completely selective, I_{max} 5.7 nmol/mg protein) for the phosphate carrier; α-cyano-4-hydroxycinnamate (non-competitive, K_I 6.3 μM) for the pyruvate carrier; phenylsuccinate (competitive, K_I 0.71 μM, K_M for succinate 1.2 mM) for succinate transport; rotenone (irreversible, but slow, I_{max} 172 pmol/mg) for complex I; antimycin (irreversible, but slow, I_{max} 127 pmol/mg) for complex III; and azide (non-competitive, K_I 10–100 μM) or cyanide (non-competitive, K_I 0.1–10 μM) for complex IV. Use of azide or cyanide is not straightforward, as they enter mitochondria, bind to different forms of cytochrome oxidase, and have had a large range of K_I values measured under a range of conditions. Cyanide may also inhibit succinate dehydrogenase at high concentrations.

2.3.3 Estimation of control coefficients from measured elasticities

The relative control coefficients of adjoining enzymes in a pathway can be estimated from their relative elasticities to their common intermediate metabolite. Elasticities may be estimated by two main methods (1, 7):

(a) Calculation from known enzyme kinetic parameters of isolated enzyme and measured concentrations of substrate and product concentrations in the system.

(b) Calculation from measurements of overall elasticities to substrate and product in the system.

The latter approach can be exemplified if we imagine a linear pathway where the enzymes are connected by single intermediates with no feedback inhibition; so that each enzyme has only two elasticities, one to its substrate and one to its product. Addition of external effectors to perturb first the top end of the pathway, and then the bottom end of the pathway, results in two independent changes in the intermediate levels and pathway flux. The flux change through any particular enzyme on addition of one of the inhibitors is given by:

$$\frac{\delta J}{J} = \varepsilon_S \cdot \frac{\delta S}{S} + \varepsilon_P \cdot \frac{\delta P}{P}$$

where S is the concentration of substrate for the enzyme, and P is the concentration of product. With two independent changes in S, P, and J we obtain two simultaneous equations, which can be solved to give the two elasticities, for each enzyme where we can measure S and P. If we can obtain all the elasticities in the system, then we can solve for all the control coefficients using the connectivity theorem. A convenient (but not obligatory) method for calculating the control coefficients from the elasticities is the matrix method (1).

Two problems with the method are:

(a) It is necessary to have prior knowledge of how many significant elasticities an enzyme has to the intermediates of the system, or the ability to estimate the elasticities to all potential internal effectors.

(b) With two or more elasticities to estimate for each enzyme, measurement errors may dominate the results.

Thomas and Fell (8) have given methods for estimating the errors involved, and suggestions for minimizing these errors.

2.3.4 Control by proton leak

Proton and other ion leaks are often significant in bioenergetic systems (reviewed in ref. 9). Some researchers have tried to measure the control coefficient of these leaks by uncoupler titrations (e.g. ref. 5). However, this is an invalid method, because it is based on the incorrect assumption that uncouplers have the same elasticity to Δp as the endogenous proton leak (10). In the absence of specific inhibitor of leaks it is necessary to estimate the control coefficient of proton leaks by measuring relative elasticities to the protonmotive force (see *Protocol 2* and refs 10, 11).

For mitochondria in respiratory states other than state 3 the proton leak flux is significant, and thus control over respiration rate is not equivalent to control over ATP synthesis. Methods of relating control over ATP synthesis to control over respiration are given below and in ref. 11.

2.4 Overall control by blocks of enzymes

If the system under study can be divided into two or three blocks of reactions interacting only via a single intermediate (or a set of intermediates whose concentrations are interdependent) then the control exerted by these blocks of reactions may be estimated as a unit. This type of control analysis is known as the 'top-down approach' (12) or 'modular analysis' (13).

2.4.1 Overall control from overall elasticities

The analysis is based on measuring the relative elasticities of the blocks to the intermediate, from which the control coefficients of the blocks can be derived (by combining the connectivity, summation, and branching theorems). Normally the level of the intermediate is changed by adding some effector of one of the blocks (block one). By measuring the dependence of the other blocks on the level of the intermediate their 'overall' or 'group' elasticity coefficient to be the intermediate can be measured. Then using another effector of a different block (not block one) the intermediate level may be changed again and the elasticity of block one to the intermediate measured. Combining the relative elasticities and the relative rates enables all the control coefficients to be estimated (using Equation 9 of *Table 1*). The effectors may be inhibitors, or stimulators, or substrate, or products of the blocks, but they must be specific to a particular block. Where a linear pathway is being examined only one flux need be measured plus the level of the intermediate, to determine the relative flux control on either side of the intermediate. The analysis may be repeated about further intermediates to progressively localize control to smaller and smaller groups of enzymes.

The top-down approach has several advantages to the bottom-up approach to control analysis:

(a) It is experimentally relatively easy to apply to complex systems.

(b) It not only estimates control coefficients but also explains them (because the relative elasticities and kinetics cause the control distribution).

(c) The same approach can be used to analyse regulation (see below).

A disadvantage of top-down control analysis is that if interactions between the blocks are mediated by more than one intermediate (or set of dependent metabolites) the analysis becomes experimentally complex (although not unanalysable, see ref. 13); or if significant interactions are not known about spurious results may be obtained.

Protocol 2. The dependence of the mitochondrial respiratory chain, proton leak, and ATP synthesis on $\Delta\psi$, and top-down control analysis of the fluxes[a]

Equipment and reagents

- Oxygen electrode, electrode meter, thermo-statted electrode vessel, water-bath, and magnetic stirrer for vessel (see Chapter 1, Rank Brothers)[b]
- TPMP⁺ (or TPP⁺) electrode inserted through gas-tight port into oxygen electrode vessel, and connected to ion meter (see Chapter 3)[c]
- Two-channel chart recorder or computer for data acquisition
- Pipettes and tips, 10 μl Hamilton syringes
- Isolated mitochondria of high known pro-tein concentration (see ref. 6) kept on ice

- 100 mM KCl, 20 mM Hepes, 10 mM K_2HPO_4, 3 mM $MgCl_2$, 1 mM EGTA, plus 20 mM glucose, 5 mM succinate, 1 mM malate, 100 μM ADP, 5 μM rotenone, brought to pH 7.2 and 37°C[d]
- Oligomycin (1 mg/ml stock)
- Malonate (K^+ salt, 0.5 M stock)[e]
- Triphenylmethylphosphonium (TPMP) bro-mide (1 mM stock)
- Hexokinase (100 U/ml stock)

A. *Method*

1. Put 5 ml of incubation medium in oxygen electrode vessel thermostatted at 37°C and add sufficient hexokinase to obtain a maximal state 3 rate (about 0.2 U/ml).[b,f]

2. Put vessel top on and exclude air bubbles.

3. Calibrate TPMP⁺ electrode with five additions of TPMP bromide up to 5 μM (see Chapter 3).

4. Add 1 mg of mitochondrial protein/ml of incubation medium.[f]

5. Wait until oxygen electrode trace is linear and there is no further change in TPMP⁺ level (up to 5 min).[g]

6. Add 200–500 nmol malonate/ml and wait until oxygen consumption is linear and measurable (about 2 min).[f,g]

7. Repeat step 6 until respiration is mostly inhibited.[h]

8. Wash out vessel and electrodes.[i]

9. Repeat steps 1–8 but with sufficient hexokinase to give about 75%, 50%, and 25% of the state 3 rate, and again with no added hexokinase but with 1 μg oligomycin/ml added at step 1.

10. Repeat steps 1–9 on separate preparations of mitochondria sufficient times to estimate errors.[j]

B. *Calculation*

1. Measure TPMP⁺ uptake and rates of respiration from chart recorder (or computer) and plot respiration rate versus TPMP⁺ uptake (or Δp) for each hexokinase level (see *Figure 1*).

2. Fit a line or curve to those points with no malonate added, but different amounts of hexokinase (broken line in *Figure 1*); this gives the dependence of the $\Delta\psi$ producing processes (the respiratory chain) on $\Delta\psi$. The slope of this curve at each point can be used to calculate the elasticity of the $\Delta\psi$ producers to $\Delta\psi$ at each level of hexokinase. The slope must be multiplied by the respiration rate and divided by $\Delta\psi$ at the point to give the elasticity.

3. Fit a curve to the points with any particular amount of hexokinase added, but different amount of malonate (solid lines in *Figure 1*); these curves give the dependence of the $\Delta\psi$ consuming processes (phosphorylation and proton leak) on $\Delta\psi$ at each level of hexokinase. The initial slope of these curves (at zero added malonate) gives the elasticity of the $\Delta\psi$ consumers to $\Delta\psi$ at each level of hexokinase. The relative elasticities of the $\Delta\psi$ producers and consumers (at any given level of hexokinase) then gives the relative control coefficients of these processes over the respiration rate (using the connectivity theorem). And since the control coefficients must sum to one, the absolute control can be calculated.

4. Fit a curve to the points with oligomycin added (open squares in *Figure 1*); this gives the dependence of the proton leak on $\Delta\psi$. This curve can be used to calculate the leak rate for any other point by assuming that the leak rate depends on $\Delta\psi$ only. Thus the oxygen consumption coupled to phosphorylation can be calculated for any point by subtracting the estimated leak rate for that value of $\Delta\psi$ from the respiration rate. Calculate the leak rate and coupled rate for each of the points without malonate (but different amounts of hexokinase). By using the branching theorem these relative rates can be used to calculate the relative control exerted by the proton leak and phosphorylation system over respiration rate. Since the sum of the control coefficients was calculated in the last step the absolute coefficients may now be calculated.

5. Control over the rates of ATP synthesis or proton leak may be calculated similarly by estimating the elasticities of these rates to $\Delta\psi$ from the same data.

---

[a] Adapted from ref. 11.
[b] If you only have a small amount of mitochondria, use a small incubation vessel and 1 ml of incubation medium.
[c] For a small incubation vessel you will need to use a combination TPMP and reference electrode. This is constructed by removing the pH-sensitive glass bulb at the end of a small commercial combination pH and reference electrode. Fill the central tube with 100 mM KCl and 10 mM TPMP bromide, and then stick on to the end a functional TPB membrane (Chapter 3; this can be difficult). A hole will have to be drilled through the vessel top to allow insertion of the TPMP electrode while keeping it gas-tight. A notch should also be cut along the side of the top to allow a syringe needle into the vessel.

Protocol 2. *Continued*

[d] This medium can obviously be adapted according to what substrate type and concentrations, ADP and ion levels, etc. you want.

[e] With an NAD^+-linked substrate rotenone can be used instead.

[f] You will have to adjust these figures according to the type and activity of your mitochondria.

[g] Check first that oxygen consumption would remain linear for the whole course of the experiment. If not change the conditions such that it is reasonably linear (e.g. lower the temperature, increase substrate concentrations), add a single aliquot of the inhibitor at a particular time point when oxygen consumption is pseudolinear, and repeat with another incubation and inhibitor concentration.

[h] If baseline of TPMP electrode drifts during experiment, add small amount of valinomycin at end of experiment to remove all TPMP from mitochondria and thus establish where baseline is.

[i] Use ethanol (or albumin) first if you have added anything which is not water soluble, then distilled water.

[j] ΔpH is assumed to be constant during these experiments, as it has been measured to be so in rat liver mitochondria (11). In different conditions ΔpH might not be constant, and it would be necessary to measure it or hold it constant by adding nigericin (see below). Kinetic interactions between the three parts of the system other than via Δψ (or Δp) will complicate the analysis considerably, and such interactions should be tested for (see refs 11, 12).

This protocol is given in some detail because it can be used for a number of purposes apart from measuring control coefficients. For example if the protocol is repeated after the addition of some effector or with mitochondria

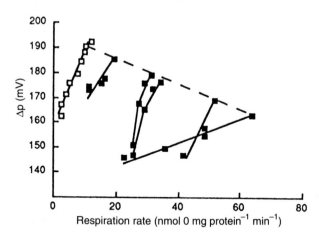

Figure 1. Determination of fluxes and elasticities of mitochondrial respiratory chain, proton leak, and phosphorylation. The solid lines are interpolations of malonate titrations of respiration rate and Δp in isolated mitochondria, at various different levels of added hexokinase to set different rates of ATP turnover. These lines represent the dependence of the rate of Δp-consuming processes on Δp. The broken line interpolates the steady states with different levels of hexokinase but no added malonate, and represents the dependence of the respiratory chain rate on Δp. The open squares are a malonate titration in the presence of oligomycin (to inhibit ATP synthesis), and represent the dependence of the proton leak on Δp. From ref. 11.

subjected to some physiological or pathological change, then the effect of this effector or change on the proton leak, respiratory chain, and phosphorylation system can be measured (e.g. the effect of thyroid hormone, see ref. 14). Or in combination with direct measurements of the rate of ATP synthesis (see Chapter 1) can be used to estimate the coupled P/O ratio as a function of $\Delta\psi$. Such experiments have shown that the coupled P/O ratio does not in fact change as a function of $\Delta\psi$ in the conditions of these experiments (15).

2.4.2 Overall control from relative flux changes

An alternative methodology to measure group control coefficients does not require the ability to measure the level of the intermediate between blocks. Only the relative flux changes of the blocks need be measured when the intermediate level is changed independently. For example in isolated mitochondria in state 3, almost all the respiratory flux is coupled to ATP synthesis via the protonmotive force. If the protonmotive force is independently decreased by adding a small amount of uncoupler the relative changes in the rate of respiration and ATP synthesis can be used to estimate the relative control exerted by these processes (16). In effect a new branch is introduced into the system at the intermediate of interest. In a branched system this is not necessary, and all that is required is the ability to specifically change each branch in turn and measure the flux changes of the other branches (12). Again it is essential that all interactions between branches or blocks are mediated by the known intermediates. If kinetic interactions are occurring between the branches mediated by effectors other than the common intermediate then spurious results may be generated, and a more complex analysis is required (13).

2.5 Control within single enzymes

Up until now we have been talking about control between enzymes, via metabolites and other effectors. But we may also be interested in control within enzymes and transporters. For example we may be interested in which step or rate constant is rate limiting within an enzyme or transporter. Metabolic control analysis can be applied in a modified form to analyse control within enzymes (17).

The extent to which a step or rate constant limits the steady state rate can be quantified as a flux control coefficient:

$$C_i = \frac{\partial J}{\partial k_i} \cdot \frac{k_i}{J}.$$

Forward rate constants have positive control coefficients, backward rate constants have negative control coefficients, and the sum of all control coefficients in an enzyme is one. Most single rate constants can not be changed individually (cf. the Haldane relation), and thus the control coefficients of forward and backward rate constants must be combined to predict real rate changes. The connectivity theorem (Equation 6 of *Table 1*) does not hold for an enzyme cycle, but there are a number of relations between the control coefficients and the enzyme intermediate levels or disequilibrium of steps (18). These relations potentially allow the control coefficients to be estimated from measurements of the enzyme intermediate levels. However, this has not been done experimentally as yet.

On the other hand if all the rate constants of an enzyme are known then the control coefficients can be calculated. This has revealed that many enzymes and transporters do not have a unique rate limiting step, and control changes dramatically with conditions (17). Control analysis of the Na^+/glucose co-transporter has revealed that control is distributed among a number of steps and changes with membrane potential, and Na^+ and glucose concentrations. Also control over the Na^+ flux differs from that over the glucose flux, and control by the individual rate constants is very high in physiological conditions.

Control coefficients can be calculated from rate constants either analytically or by modelling. If the mechanism involved is simple the rate equation (expressing the rate in terms of the rate constants and substrate and product concentrations) it can be looked up in a book of enzyme kinetics. This equation is then differentiated with respect to each rate constant, and multiplied by the rate constant divided by the rate (see definition above). For more complex mechanisms it is simpler to use computer modelling to numerically calculate the steady states and determine the effect of changing rate constants. For example, SCAMP (19) and MetaModel (20) are metabolic modelling programs which calculates all the control coefficients and can be used for single enzyme cycles.

3. The analysis of regulation

Regulation is a rather vague term as normally used, and thus it is necessary to clarify its meaning. 'How is a process regulated?' usually means 'how are physiological changes in the rate of the process brought about?'. In part this is a non-quantitative question about what metabolic or signal-transduction effectors and pathways are involved in mediating between the supracellular events and intracellular molecular changes in the process itself. But there are also quantitative questions such as 'to what extent is the physiological response brought about via a particular effector or pathway?' or 'to what extent does a particular effector change a pathway rate via direct effects on the individual enzymes of the pathway?' Control and regulation of mitochondrial respiration and ATP synthesis are reviewed in ref. 21.

3.1 Regulation analysis within control theory

In metabolic control theory the response coefficient quantifies the sensitivity of pathway rate to an external effector (see above). Thus if the effector is a physiological regulator, then its response coefficient quantifies the sensitivity of the system flux to the level of the regulator in the particular conditions studied, and thus it quantifies the potential of an effector to regulate a flux. Whether or not this potential is realized depends on the concentration change of the effector physiologically: the fractional concentration change multiplied by the response coefficient equals the fractional flux change (for small changes).

The response is made up of two components: first the sensitivity of the 'isolated' enzyme to the effector (i.e. the elasticity of the enzyme to the effector), and second the sensitivity of the pathway flux to the 'activity' of the enzyme (i.e. the flux control coefficient of the enzyme). The response coefficient is equal to elasticity coefficient multiplied by the control coefficient (Equations 4 and 8 of *Table 1*). Thus measuring the elasticity and control coefficients predicts and explains the response coefficient.

However, the response may be due to the effector altering the kinetics of more than one enzyme in the pathway, that is the combined response may be sum of a number of partial (individual) responses mediated by different enzymes. And each partial response coefficient is the product of the elasticity of the enzyme to the effector and the control coefficient of the enzyme over the pathway (see Equation 8 of *Table 1*). Thus the combined response coefficient of a pathway to an effector can be expressed as a sum of a number of partial responses due to individual enzymes, and these partial responses quantify the degree to which the total (combined) response is mediated by the individual enzymes. Thus measuring the partial responses is a way of quantifying the extent to which a response is brought about by different mechanisms.

Other ways of quantifying regulation within metabolic control theory are discussed in ref. 22. However, control theory deals with differentials, and thus can describe and predict the effects of small changes only. On the other hand, in a linear pathway if the control coefficient of an enzyme is measured to be close to zero, then it can be predicted that no amount of activation of that enzyme will significantly affect the pathway flux. And similarly if the control coefficient is measured to be close to one, then any amount of inhibition of that enzyme will leave the control coefficient of the enzyme at close to one. General methods for analysing large changes are outlined below.

3.2 Cross-over

One of the traditional methods for determining where in a pathway an effector acts is the 'cross-over' method. The theory of the method is as follows. If an effector inhibits a linear pathway, with no feedback or feed-forward loops, and acts at only one position in the pathway, then all the pathway intermediates prior to the point of inhibition should rise in level and all the intermediates after the point of inhibition should fall in level. Thus

Guy C. Brown

the point of inhibition can be localized to the enzyme between the last intermediate to rise and the first intermediate to fall.

However, this logic fails to apply if the effector acts at more than one enzyme in the pathway or there are feedback or feedforward loops in the pathway. Cross-over analysis itself can not detect whether such complexity is present, and thus the results of such an analysis are not conclusive. Further analysis requires other methods.

3.3 Phenomenological kinetics

Phenomenological kinetics is just the kinetics as measured, in the absence of any theory. That is the measured dependence of the rate of a process on its substrates, products, and/or other effectors. As such phenomenological kinetics constitutes the bulk of what is published on experimental 'control and regulation in bioenergetics'; for example the dependence of the rate of mitochondrial respiration or a reconstituted transporter on the membrane potential. But what can we do with the phenomenological kinetics of a process?

There are at least three uses.

1. We can try and look inside the process, and see how the phenomenological kinetics are produced from the behaviour of subcomponents of the system. For example enzyme kinetics tries to explain the phenomenological kinetics of isolated enzymes in terms of kinetics of subreactions within the enzyme. And computer modelling of metabolic systems tries to explain phenomenological behaviour of pathways by combining the known behaviour of the individual component enzymes and transporters. A variety of programs (19, 20) are now available for modelling metabolic or enzyme processes, and do not require the user to have any knowledge of computer programing or modelling. The program just asks you for the structure of the system, and the equations for the components, and it then calculates the steady states, transitions, control properties, etc.

2. We can combine the measured phenomenological kinetics of different systems to predict the behaviour of larger sized systems. For example having measured the dependence of mitochondrial respiratory chain, proton leak, and phosphorylation system on their intermediate (membrane potential) in *Protocol 2*, we can combine this information to predict what will happen to the rates of these three processes if the effective activity of one of the processes is doubled. Thus the phenomenological kinetics can be used to predict the behaviour of larger sized systems.

3. We can compare the kinetics before and after some change, in order to investigate where within a system the change has directly acted. For example, if we repeat *Protocol 2* after having added some effector (e.g. a hormone or ion) to the mitochondria, then we can determine whether and how the effector directly changes the kinetics of the respiratory chain, proton leak, or

phosphorylation system. This approach to determining where within a system an effector acts overcomes many of the limitations of the cross-over method. It is sometimes called 'phenomenological kinetic analysis' or 'elasticity analysis', but it has been used extensively in the past, and is really an extension of cross-over analysis. The approach has been used in bioenergetics, for example to determine how thyroid hormone changes mitochondrial respiration (14), and whether the stimulation of respiration by calcium-mobilizing hormones is entirely due to stimulation of NADH production (23).

3.3.1 Regulation by membrane potential and ΔpH

A relatively common problem encountered in bioenergetics is to determine the effect of $\Delta\psi$ or ΔpH on a process, for example on a rate of transport. This requires the ability to manipulate the levels of $\Delta\psi$, ΔpH, or Δp.

A high steady state level of Δp can be produced by adding excess substrate for the proton pumps (respiratory substrate, light, or ATP). Steady states with a range of lower values of Δp can then be produced by titrating either with an uncoupler or inhibitors of the proton pumps (e.g. respiratory chain inhibitors or oligomycin; see *Protocol 2*).

$\Delta\psi$ and ΔpH can be altered differentially using valinomycin or nigericin in mitochondria, bacteria, and vesicles. Valinomycin is a K^+ ionophore, catalysing electrogenic K^+ uniport of high selectivity. Nigericin is a K^+/H^+ ionophore, catalysing electroneutral exchange of one H^+ for one K^+ ion. Addition of 'excess' valinomycin will clamp $\Delta\psi$ equal to the K^+ gradient. Addition of nigericin will clamp ΔpH equal to the K^+ gradient. A range of ΔpH values, with an almost constant $\Delta\psi$, can be obtained by adding valinomycin (with a low but significant amount of external K^+), and then inhibitor titrating the proton pumps, or titrating with uncoupler, to obtain a range of steady states with different values of ΔpH. Similarly to obtain a range of $\Delta\psi$ values, with an almost constant ΔpH, nigericin should be added (with a high level of external K^+), and then inhibitor titrate the proton pumps, or titrate with uncoupler.

Alternatively the external K^+ concentration can be changed to set a range of different $\Delta\psi$ values (with valinomycin) or ΔpH values (with nigericin). This can also be done when the proton pumps are not operating, or are not present (for example in vesicles). In the absence of active proton transport, the proton leak will cause Δp to return to zero after valinomycin (or nigericin) addition, so that the steady state $\Delta\psi$ value will be balanced by an equal and opposite ΔpH.

The main problems with using valinomycin and nigericin are:

(a) The K^+ movements continuously change the K^+ gradient and thus the $\Delta\psi$ and ΔpH values.

(b) The K^+ and other ion movements cause the enclosed volume to change which often causes secondary effects on transport.

Changes in enclosed volume due to ion movements cause internal concentrations to change continuously, and this may effect the activity of transporters and internal enzymes (e.g. mitochondrial respiration is inhibited by contraction). Swelling causes stretch activation of some transporters, and eventually ruptures membranes (small vesicles being ruptured first). The change in K^+ concentrations (for any given rate of K^+ movement) can be minimized by having relatively high external and internal levels of K^+. However this gives access to only relatively low $\Delta\psi$, values (with valinomycin). On the other hand, when the proton pumps are also operating to set a high value of $\Delta\psi$, valinomycin and potassium can be used to clamp high values of $\Delta\psi$, with relatively small K^+ movements. K^+ movements can also be minimized, when the proton pumps are operating, by using very small ('limiting' rather than 'excess') concentrations of valinomycin or nigericin. This alters the previous steady state of the endogenous K^+ uniport and K^+/H^+ antiport activities to give a new steady state with a new value of $\Delta\psi$ and ΔpH, but requires no net transport of K^+ in the steady state. At high concentrations valinomycin also has a significant proton uniport activity.

References

1. Fell, D. A. (1992). *Biochem. J.*, **286**, 313.
2. Kholodenko, B., Zilinskiene, V., Borutaite, V., Ivanoviene, L., Toleikis, A., and Praskevicius, A. (1987). *FEBS Lett.*, **223**, 247.
3. Small, J. R. (1993). *Biochem. J.*, **296**, 423.
4. Groen, A., van der Meer, R., Westerhoff, H., Wanders, R., Akerboom, T., and Tager, J. (1982). In *Metabolic compartmentation* (ed. S. Sies), pp. 9–37. Academic Press, London.
5. Groen, A. K., Wanders, R. J. A., Westerhoff, H. V., van der Meer, R., and Tager, J. M. (1982). *J. Biol. Chem.*, **257**, 2754.
6. Rickwood, D., Wilson, M. T., and Darley-Usmar, V. M. (1987). In *Mitochondria: a practical approach* (ed. V. M. Darley-Usmar, D. Rickwood, and M. T. Wilson), pp. 1–16. IRL Press, Oxford.
7. Groen, A. K., van Roermund, C. W. T., Vervoorn, R. C., and Tager, J. M. (1986). *Biochem. J.*, **237**, 379.
8. Thomas, S. and Fell, D. A. (1994). *J. Theor. Biol.*, **167**, 175.
9. Brand, M. D., Hafner, R. P., and Brown, G. C. (1988). *Biochem. J.*, **255**, 535.
10. Brown, G. C. (1992). *FASEB J.*, **6**, 2961.
11. Hafner, R. P., Brown, G. C., and Brand, M. D. (1990). *Eur. J. Biochem.*, **188**, 313.
12. Brown, G. C., Hafner, R. P., and Brand, M. D. (1990). *Eur. J. Biochem.*, **188**, 321.
13. Schuster, S., Kahn, D., and Westerhoff, H. V. (1993). *Biophys. Chem.*, **48**, 1.
14. Hafner, R. P., Brown, G. C., and Brand, M. D. (1990). *Biochem. J.*, **265**, 731.
15. Hafner, R. P. and Brand, M. D. (1991). *Biochem. J.*, **275**, 75.
16. Westerhoff, H. V., Groen, A. K., and Wanders, R. J. A. (1983). *Biochem. Soc. Trans.*, **11**, 90.

17. Brown, G. C. and Cooper, C. E. (1993). *Biochem. J.*, **294**, 87.
18. Brown, G. C. and Cooper, C. E. (1994). *Biochem. J.*, **300**, 159.
19. Sauro, H. M. (1993). *Comp. Appl. Biosci.*, **9**, 441.
20. Cornish-Bowden, A. and Hofmeyr, J.-H. S. (1991). *Comp. Appl. Biosci.*, **7**, 89.
21. Brown, G. C. (1992). *Biochem. J.*, **284**, 1.
22. Hofmeyer, H. S. and Cornish-Bowden, A. (1991). *Eur. J. Biochem.*, **200**, 223.
23. Brown, G. C., Lakin-Thomas, P. L., and Brand, M. D. (1990). *Eur. J. Biochem.*, **192**, 355.

7

Patch-clamping of mitochondrial membranes and of proteoliposomes

CRISTINA BALLARIN and M. CATIA SORGATO

1. Introduction

The electrophysiological analysis of mitochondrial membranes developed in recent years has shown the presence of several high conductance ion channels (1). The physiological roles played by these channels are not yet clearly established. However, the continuous increase of information and the contributions of different expertise render the topic an exciting focus of interest.

This chapter describes the essential steps to accomplish the physiological characterization of mitochondrial proteins at a single channel level with the use of the patch-clamp technique applied directly to the native membranes of the organelle. Integral mitochondria must be used in order to study the outer membrane *in situ*. This approach has proved difficult in view of the small dimensions of the integral organelle. However, if the outer membrane is removed (yielding the so-called mitoplast), then the inner membrane expands sufficiently to facilitate its analysis with a patch-clamp electrode. The mitochondria most commonly used for this purpose are from mammalian tissues (1). More recently yeast mitochondria have also been subjected to the patch-clamp technique (C. Ballarin and M. C. Sorgato, in preparation), and thus protocols are provided for isolating either type of mitochondria and for preparing mitoplasts. Sometimes the use of proteoliposomes in electrophysiology may be advantageous (1). Therefore a method is described for adequately enlarging liposomes containing isolated mitochondrial membranes, or some fraction of them.

The patch-clamp is a versatile and extremely sensitive technique which has now become very popular, and many of the available papers will provide the reader with competent and detailed descriptions of the electronics and operational modes of the system as well as computer-based single channel analysis (2, 3). (For the structure and function of ion channels, we strongly recommend ref. 4.) In view of the practical aim of this chapter, we will illustrate the steps needed to accomplish the patch-clamping of mitochondrial membranes in sufficient detail to allow its repetition. It should, however, be borne in mind that successful experiments with the patch-clamp technique depend on several, independent parameters, the complete control of which can only be achieved with experience.

2. Preparation of large mitochondria and mitoplasts from mice livers

The purpose of this procedure is to obtain a population of large mitochondria, and then of large mitoplasts, from tissues of experimental animals in order to have vesicles large enough to be amenable to the patch-clamp electrode. Undoubtedly, homogenization of the tissue is the key and most delicate step; it must be sufficiently intense to break the cell membrane but at the same time gentle enough to leave the bigger mitochondria intact. It is therefore best to use livers with a soft texture (such as those from mice) and a loose-fitting glass ball homogenizer. If this procedure is followed, it is easily possible to obtain mitochondria that are approximately 1–1.5 μm in diameter in a fairly high yield, and mitoplasts with a mean diameter of around 3 μm. Alternatively, unweaned 17–19 day-old mice can be fed with deionized water and with an inducer of giant mitochondria in the liver, cuprizone (oxalic acid bis(cyclohexylidenehydrazide)) (Aldrich-Chemie; G. F. Smith Chemicals) (5), at a concentration of 3 g per 500 g of ground rodent chow. According to the literature, mitochondria with a diameter as large as 5–10 μm are present in the liver after three to four days of this diet, but in our experience such huge organelles are not easy to isolate. The presence or absence of cuprizone in the diet does not change the electric behaviour of mitochondrial membranes (1). However, given their bigger size, and if the outer membrane is to be analysed, or if patch-clampers are inexperienced, cuprizone mitochondria and mitoplasts are probably more suitable.

2.1 Preparation of mitochondria

The method given below *(Protocol 1)* is equally applicable to livers of cuprizone treated and untreated animals.

Protocol 1. Preparation of large mitochondria from mice liver[a]

Equipment and reagents

- Refrigerated superspeed centrifuge
- Refrigerated swing-out bench centrifuge
- 7 ml homogenizer with a glass ball pestle (clearance of 0.006–0.009 cm)
- Isolation medium: 200 mM mannitol, 70 mM sucrose, 2 mM Hepes (pH 7.4 with

KOH), 0.5 mg/ml BSA (added immediately prior to use)[b]
- Suspending medium: 250 mM mannitol, 50 mM Hepes (pH 7.2 with KOH)
- 0.5 M sucrose

Method

1. Obtain one or two livers from mice killed by cervical fracture and place them in a 25 ml beaker containing 10 ml of cold isolation medium. Mince the tissue finely with sharp scissors. Allow time for the tissue to settle, then decant, and resuspend in fresh medium. Repeat this operation three to four times, or until there is no visible trace of blood, fat, or connective tissue. Accomplish the entire step rapidly.

2. Transfer the tissue into the homogenizer using an amount of tissue sufficient to occupy approximately half of its volume, and fill the rest with the isolation buffer. Insert the loose glass ball pestle and gently rotate it until it reaches the bottom. Two such strokes are sufficient.

3. Pellet[c] the homogenate at 120 g for 1 min using thick glass tubes (10 ml) inserted into rubber adaptors. Gently remove and filter supernatants through two layers of cheese cloth. Save them.

4. Recover any large mitochondria still present with the cell debris, by homogenizing the soft pellets as before (each with 5 ml of the isolation medium). Filter and centrifuge again as in step 3. Pool the supernatants with those saved previously.

5. Lay the pooled supernatants on top of cold 0.5 M sucrose contained in 10 ml plastic test-tubes (5 ml of supernatant per 5 ml of sucrose), and pellet for 10 min at 730 g, using a refrigerated swing-out bench centrifuge (corresponding to 2260 r.p.m., using a Heraeus Omnifuge 2.0RS with a 3360 rotor; r = 12.8 cm).[c]

6. Discard the top layer, containing the smaller mitochondria, up to approximately 1 cm above the sucrose phase. This part frequently (though not always) looks turbid and may contain large mitochondria. Using a Pasteur pipette, transfer this part into a graduated cylinder along with the cloudy sucrose phase, paying attention not to remove the loose, fluffy pellet and the blood spot.

7. Dilute the saved volume, consisting mainly of 0.5 M sucrose, to 0.3 M sucrose by adding slowly the required volume of cold bidistilled water whilst stirring, and then centrifuge for 5 min at 750 g using the 10 ml glass test-tubes.

8. Discard the supernatant. Dissolve the hard, yellowish pellet in a few drops of isolation medium and then dilute to 5 ml with the same buffer. If a blood spot is still present, take care not to disturb it.

9. After laying the resuspended material on top of 5 ml of 0.5 M sucrose (with v/v of 1), centrifuge for 3 min at 410 g (corresponding to 2080 r.p.m., using a Heraeus Omnifuge 2.0RS with a 3360 rotor; r = 8.5 cm) in a swing-out bench centrifuge.[c] Save the upper layer, including the interface.

Protocol 1. *Continued*

10. Centrifuge as in step 7 and dissolve the pellet (containing large mitochondria) in 1–2 ml of the suspending medium.

[a] All operations are carried out at 4°C using cold and filtered solutions (see section 6.3).
[b] Occasionally, millimolar concentrations of ethylene glycol-bis-(b-aminoethyl ether) N,N'-tetraacetic acid or ethylenediaminetetraacetic acid have been added to the isolation medium. However, the electric pattern of mitochondrial membranes studied by us remained unaltered (1), so that the presence of calcium chelators appears to be only a matter of personal choice.
[c] Centrifugations should be allowed to decelerate without the use of automatic brakes.

2.2 Preparation of mitoplasts

Mitoplasts are mitochondria without the outer membrane. Removal of the outer membrane allows expansion of the invaginated inner membrane yielding a vesicle with a larger diameter than that of the parent mitochondria. In this way, vesicles are better resolved under the optical microscope and the inner membrane can be directly studied with the patch-clamp. The outer membrane can be broken by osmotic shock, as detailed in *Protocol 2*. Frequently though, a piece of the outer membrane remains attached to the inner one through tight junctions, giving rise to the so-called cap region (1) (see *Figure 1a*).

Protocol 2. Preparation of large mitoplasts from mice liver[a]

Equipment and reagents

- Refrigerated superspeed centrifuge
- Swelling medium: 10 mM Tris (pH 7.5 with H₃PO₄)
- Shrinking medium: 1.8 M sucrose, 2 mM ATP, 2 mM MgSO₄
- Suspending medium: 250 mM mannitol, 50 mM Hepes (pH 7.2 with KOH)

Method

1. First dissolve the mitochondrial pellet (see *Protocol 1*) in 50–100 µl of the suspending medium, and transfer it into a 30 ml thick glass tube. Then slowly dilute to 15 ml with the swelling buffer. Let the tube stand on ice for 7 min (or 5 min if cuprizone mitochondria are used) with occasional stirring.

2. Very slowly add 5 ml of the shrinking buffer, stir, and leave for a further 5 min on ice. The suspension becomes slightly turbid.

3. Using the tubes protected by rubber adaptors, centrifuge for 5 min at 4300 g.[b]

4. Suspend the very small, light coloured pellet in a few (1–5) ml of the suspending medium, and after transferring it into the thick glass 10 ml test-tube, wash the mitoplasts at 4300 g for 5 min.[b]

5. Dissolve the resulting pellet in the suspending medium. In general, a soft pellet of around 0.5 cm in diameter is obtained from one liver.

[a] All operations are carried out at 4°C using cold and filtered solutions (see section 6.3).
[b] Allow centrifugations to decelerate without the use of automatic brakes.

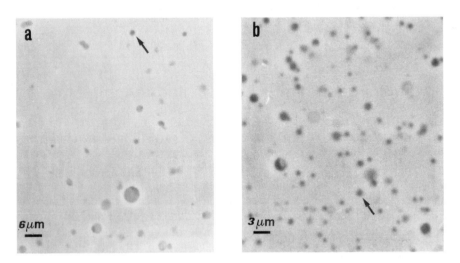

Figure 1. Vesicles observed at the patch-clamp microscope after subjecting mice liver (a) and yeast (b) mitochondria to a swelling step (see *Protocols 2* and *3* and section 3.3). Note the bigger size with respect to parent mitochondria (*arrows*) and the presence of cap regions (*dark zones*) at the periphery.

A faster way to obtain mitoplasts, albeit with a lower yield, involves swelling the mitochondria directly in the patch-clamp dish (see *Protocol 3*).

Protocol 3. Preparation of mitoplasts directly on the patch-clamp dish

1. Place 1–5 μl of the mitochondrial suspension (at a concentration of around 10–30 mg/ml) in the centre of the glass part of the dish (see *Protocol 12*).

2. Cover the small drop with approximately 1 ml of the swelling medium (as detailed in *Protocol 12*, step 2). Allow at least 3 min for the vesicles to sediment and adhere to the bottom of the dish, and then follow the formation of the mitoplasts directly under the microscope at a magnification of at least ×500.

Protocol 3. *Continued*

3. When the vesicles augment their size, and bear a darker zone (cap region) usually concentrated in a peripheral part of the vesicle (see section 2.2 and *Figure 1a*), carry out extensive perfusion with the experimental buffer, either using the perfusion set-up (see *Figure 2*), or manually by withdrawing and replacing the swelling medium with two syringes (see section 6.3), the tips being placed at the border of the dish.

3. Preparation of mitochondria and mitoplasts from yeast

The electric characterization of membranes from yeast mitochondria by the patch-clamp technique is desirable for at least two reasons. One is to ascertain whether an identical channel activity is present in the mitochondria of different eukaryotic cells (mammalian and yeast) which are phylogenetically distant but apparently share similar physiology. The second reason is that yeast genomes can be manipulated relatively easily. The use of appropriate mutants can therefore help to shed light on the role of these channels and also to isolate the proteins responsible for these activities. On the other hand, a disadvantage of yeast mitochondria is their minute size, at the limit of the resolution of optical microscopes, which hampers the patch-clamping of the integral organelle. This obstacle can, however, be overcome by the use of proteoliposomes containing both inner and outer membranes or the isolated outer one. Conversely, mitoplasts of sufficient dimensions can be obtained (see *Figure 1b*), so that the study of the inner membrane can be carried out with confidence.

3.1 Growth and harvesting of yeast cells

As standard methods for the storage and revival of yeast strains can be found in any laboratory manual (as, for example, ref. 6), the protocols given below for cell growth and harvesting (*Protocol 4*) and preparation of mitochondria (*Protocol 5*) start from the growth of cells.

Protocol 4. Growth and harvesting of yeast cells

Equipment and reagents

- Shaking incubator kept at 30°C
- Sterile 3 litre Erlenmeyer flask
- Superspeed centrifuge with a rotor which accommodates 250 ml bottles
- Sterile YP medium: 1% yeast extract, 2% Bactopeptone, 3% glycerol, pH 5 (glycerol is usually added immediately prior to use from a 60% sterile stock solution)

Method

1. Inoculate 3–6 ml of the yeast starter culture[a] into 300 ml YP medium (contained in a 3 litre Erlenmeyer flask) and incubate at 30°C with vigorous shaking, until it reaches an OD_{600nm} of between 1 and 2 (after around 15–20 h).

2. Centrifuge[b] the culture for 5 min at 1460 g, using two 250 ml bottles.

3. Discard the supernatants and resuspend the pellets by gentle shaking in few millilitres of bidistilled water. After determining the weight of an empty 250 ml bottle, transfer the pellets into it. Fill the bottle with more bidistilled water.

4. Centrifuge for 5 min as in step 2, and remove the supernatant. Determine the wet weight of the cells by subtracting the bottle weight from the bottle plus pellet weight. At most, 2–3 g of cells are usually obtained.

[a] A starter culture consists of 20–50 ml culture grown until stationary phase. It can be used within two weeks.
[b] All centrifugations of this protocol can be carried out at room temperature.

3.2 Isolation of mitochondria from yeast

See *Protocol 5*.

Protocol 5. Preparation of yeast mitochondria

Equipment and reagents

- Refrigerated superspeed centrifuge with a rotor accommodating 10 ml and 30 ml tubes
- 30 ml homogenizer with a loose glass ball pestle
- 25 ml Erlenmeyer flask
- DTT medium: 100 mM Tris–SO_4, 10 mM dithiothreitol, pH 9.4
- 1.2 M sorbitol
- Zymolyase medium: 1.2 M sorbitol, 20 mM K-phosphate, 6.5 mg zymolyase (20T) (Seikagaku Co.; ICN Biochemicals) per gram of cell wet weight, pH 7.4 (the buffer

is usually prepared immediately prior to use by dilution of the stock solutions; the zymolyase powder is then added to the buffer after being dissolved in few drops of 30% glycerol).
- Homogenization medium: 0.6 M sorbitol, 10 mM Tris, 1 mM ethylenediaminetetra-acetic acid, 0.2% BSA fatty acid-free, 1 mM phenylmethylsulfonyl fluoride (Sigma), pH 7.4
- Suspending medium: 0.6 M mannitol, 10 mM Tris–HCl pH 7.4

Method

1. Suspend the pellet (obtained in *Protocol 4*, step 4) using 2 ml of DTT medium per gram of cell wet weight, and transfer the material into a 25 ml Erlenmeyer flask.

2. Shake the flask for 15–20 min in the incubator at 30°C with sufficient force to prevent sedimentation of the cells.

Cristina Ballarin and M. Catia Sorgato

Protocol 5. *Continued*

3. Centrifuge[a] the material for 5 min at 1930 *g*, using 10 ml glass, or plastic, tubes.

4. Suspend the pellet in 10 ml of 1.2 M sorbitol and centrifuge as in step 3.

5. Suspend the pellet in the zymolyase medium and incubate for 20–60 min at 30°C as in step 2. A time span is given for the incubation period to obtain the spheroplasts because for each incubation it is necessary (starting after 20 min, then every 5–10 min) to follow the action of the enzyme on the cell wall. This can be done by eye,[b] by diluting 30 μl of the cell suspension in 3 ml of water:

   (a) If the water remains clear, cell lysis has occurred, thus indicating the completion of the cell wall digestion.

   (b) Otherwise the water becomes slightly turbid: in this case further incubation is necessary. If, however, no lysis occurs after 1 h, prepare a new culture.

6. Centrifuge the suspension for 5 min as in step 3.

7. Suspend the pellet in at least 20 ml of 1.2 M sorbitol and centrifuge as in step 3.

8. Suspend the pellet in the homogenization medium (1 ml per 0.15 g of cell wet weight) and after transfer to the homogenizer, gently homogenize the spheroplasts 10–15 times without rotation of the pestle.

9. Transfer the homogenate into a 30 ml plastic centrifuge tube. Rinse the homogenizer with the same volume of the homogenization medium used in step 8, and add it to the homogenate. Centrifuge for 10 min at 480 *g*.

10. Centrifuge the supernatant as in step 9.

11. Centrifuge the supernatant for 10 min at 3000 *g*. The pellet, containing mitochondria, is suspended in 50–100 μl of the suspending medium (at a final concentration of approximately 10 mg/ml).

[a] All centrifugations on this protocol must be run at 4°C.
[b] Alternatively, the formation of spheroplasts can be followed spectrophotometrically: a 50 times reduction in the suspension's OD_{600nm} compared to that at time zero means that complete cell wall digestion has been accomplished.

3.3 Preparation of yeast mitoplasts

We found that the procedure used for preparing mammalian mitoplasts (see *Protocol 2*) completely disrupted yeast mitochondria, most probably because of the too high difference in osmolarity between the swelling medium (of around 20 mOsm) and the yeast organelle (of around 600 mOsm). Conversely, mitoplasts, closely resembling mammalian ones (see *Figure 1a*), were

140

obtained by incubating mitochondria in 110 mOsm as described in ref. 7. However, with this preparation good seals could be formed only within a very limited period of time, possibly because the inner membrane was rendered fragile by the prolonged osmotic shock (7). On the other hand, if a few microlitres of the mitochondrial stock suspension were diluted directly in the patch-clamp dish with the experimental medium [composed of 150 mM KCl, 20 mM Hepes (pH 7.2 with KOH), 0.1 mM $CaCl_2$; 350 mOsm], then mitoplasts (see *Figure 1b*) formed in approximately 5–10 min with a much lower yield but with a slightly more resistant membrane. For this reason, we generally preferred to patch-clamp the latter preparation. Importantly, the electric activity of the inner membrane was independent of the method used to prepare the vesicles (C. Ballarin and M. C. Sorgato, in preparation).

Except for the medium used, steps to follow for obtaining mitoplasts are as reported in *Protocol 3*.

4. Giant proteoliposomes

If mitochondria or mitoplasts can not be resolved satisfactorily with an optical microscope, or if one wishes to test the electric activity of a particular membrane fraction, then it is necessary to incorporate the membranes into liposomes (see *Protocol 6*). Subsequently, the proteoliposomes have to be suitably enlarged using, for example, a delicate dehydration–rehydration procedure (see *Protocol 7*). However, the high protein content typical of the inner membrane may often be an obstacle for successful experiments in terms of the rate of incorporation into liposomes and/or of detection of the desired channel activity. It is therefore advisable to use mitochondrial membranes already diluted with exogenous lipids (such as asolectin liposomes, see section 4.1).

4.1 Preparation of small liposomes and dilution of mitochondrial membranes

Any of the protocols reported in the literature can be adopted for preparing asolectin liposomes. Our preferred method (with a starting concentration of 100 mg lipid/ml) is described in ref. 8, with the additional purification of lipids carried out according to refs 9 and 10. Liposomes are then stored at −80 °C as 100–200 µl aliquots. The procedure of lipid enrichment of mitochondrial membranes differs slightly depending upon whether the starting material is the isolated membrane or whole mitochondria.

Protocol 6. Lipid enrichment

Equipment and reagents

- Airfuge
- Desiccator
- Microscope with at least 100× magnification
- 10 mM Mops-KOH pH 7.2 with and without 5% (v/v) ethylene glycol

---

Protocol 6. *Continued*

A. *Isolated inner or outer mitochondrial membrane[a]*

1. Mix 5–20 μl of the membrane fraction (containing 50–200 μg total protein[b]) with 180 μl of asolectin liposomes.

2. Pellet the mixture using a Beckman Airfuge run at the maximum speed.

3. Remove carefully the supernatant with a syringe. Subject the pellet to the enlargement step (see *Protocol 7*) or suspend it in a few microlitres of 10 mM Mops-KOH pH 7.2 and then store at −80°C.

B. *Whole mitochondria*

1. Mix 10–40 μl of the mitochondrial suspension (containing 50–200 μg total protein[b]) with 200 μl of liposomes (see section 4.1) in a 50 ml round-bottomed flask.

2. Freeze the mixture by rapidly rotating the flask in a dry ice–ethanol bath. Allow the material to thaw at room temperature, then repeat the freeze–thaw step twice more. Use the material (see *Protocol 7*) or store it as 10–50 μl aliquots at −80°C until use.

[a] One may choose any of the methods available in the literature to isolate the inner and outer membrane, for example that described in ref. 11.
[b] It is important to stress that the given protein concentrations are approximations. The optimal concentration to use depends upon the activity of the channel protein one wishes to analyse and on the membrane fraction available. Therefore the extent of the lipid enrichment has to be adjusted out of experience. For example, mitochondrial fractions containing both membranes prepared according to *Protocol 6B* are frequently passed through a second lipid enrichment step identical to *Protocol 6A*.

Protocol 7. Enlargement of small proteoliposomes

Equipment and reagents

• See *Protocol 6*

Method

1. Using the tip of an automatic pipette, mix fresh or stored mitochondrial fractions (see *Protocol 6*) with 20–30 μl of 10 mM Mops-KOH pH 7.2 plus 5% (v/v) ethylene glycol (which prevents excessive dehydration).

2. Deposit the suspension as a drop into the centre of a glass slide, then place the slide in a desiccator containing $CaCl_2$ granules, at 4°C, until the drop has been reduced to a translucent film.[a]

3. Remove the slide from the desiccator and rehydrate delicately with 20 μl of the buffer to be used in the patch-clamp experiment. Add the

solution by starting from the rim of the drop and working inwards until the partially dehydrated spot is completely covered.

4. Deposit the slide on a paper pad, which has been wet with water, contained in a Petri dish. Close the dish tightly and leave it at 4°C for at least 3–4 h.[b]

5. Observe (under a ×100 magnification) the multi-and unilamellar proteoliposomes with diameters ranging from 2–10 μm (but sometimes higher) formed at the edge of the drop. To obtain the vesicles free of the amorphous material present in the centre of the drop, firstly loosen them from the edge by directing the flow of a few microlitres of the medium (to be used in the patch-clamp experiment) away from the drop and then collect the vesicles with a syringe. Sometimes, however, if the rim is sufficiently clean, you may collect the proteoliposomes directly from the spot.[c]

[a] The dehydration process usually takes between 3–5 h. Avoid excess dehydration.
[b] For convenience one can allow the rehydration process to run overnight.
[c] If no vesicles are found, the most probable cause is that too high a protein concentration was used.

5. The patch-clamp set-up

A good set-up must satisfy the essential requirements for measuring the electric activity of any biological membrane. These are the mechanical stability and the amplification of very small current signals (in the pA range) with low noise. The various parts that form a patch-clamp set-up (schematically drawn in *Figure 2*) are now briefly described.

5.1 Mechanical set-up

An important piece of equipment is the anti-vibrational table, made from a heavy slab (marble, for example) on pneumatic supports. The table nullifies mechanical vibrations deleterious for the micromanipulator-driven micrometre movements of the pipette necessary to reach the membrane and therefore to make a seal. It also preserves the stability of the patch itself, especially when in the cell-attached configuration. The Faraday cage covering the table is indispensable for protecting the headstage of the amplifier (see section 5.2) from extraneous electric pick-up. Furthermore, it is useful to mount a movable metal net or a conductive plastic roll over the front opening of the cage; this must be lowered during the recordings and can be lifted easily when necessary. The other constituents, placed on the table, are the metallic tower and the inverted microscope.

The tower serves as a support for the mechanical macromanipulators connected to the hydraulic (or piezoelectric) micromanipulators (List-electronic;

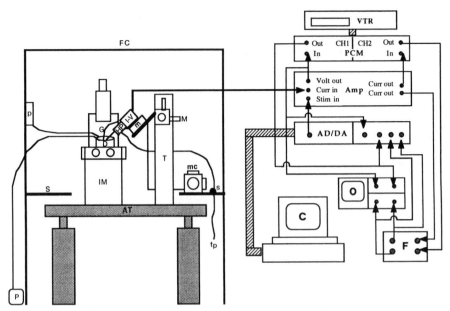

Figure 2. The patch-clamp set-up. Mechanical set-up: AT, anti-vibrational table; S, support isolated from AT; FC, Faraday cage; IM, inverted microscope; D, experimental dish; T, tower; M, macromanipulator; m, micromanipulator; mc, micromanipulator remote control; I-V, headstage (I-V converter); HP, holder and pipette; G, ground electrode; P, peristaltic pump; p, syringe with the perfusing solution; tp, tube for pressure control; s, switch to hold the pressure. Electronic set-up: VTR, video tape recorder; PCM, pulse code modulator; Amp, amplifier; O, oscilloscope; AD/DA, analog to digital/digital to analog converter; F, filter; C, personal computer. An easy to build perfusion set-up (p, P) consists of a syringe (p) (attached to the Faraday cage) and of a peristaltic pump (P) (placed outside the cage), connected to the dish through plastic tubes for the inflow and outflow of solutions. A simple way to impose a pressure (positive or negative) to the pipette consists of connecting one end of a plastic tube (tp) to the holder and by blowing or sucking by mouth at the other end. The tube is passed through a switch (s) which can be used to hold the pressure constant. For other details see section 5 and text.

Narishige). Attached directly to the latter is the headstage of the amplifier which in turn accommodates the pipette holder. It is very important to mount all the various pieces in such a way to position the pipette holder at an angle of approximately 45° with respect to the mechanical stage of the microscope, where the experimental dish will be placed. Apparently, such inclination is required for the pipette to touch the membrane of a spherical vesicles in the correct way (see *Protocol 13*). The direct handling of the macromanipulators drives the coarse movements of the pipette along the three axes, and is utilized to focus the tip and to bring it close to the vesicle. The final 3-D movements of the pipette needed to make a seal are accomplished by means of the micromanipulators driven by remote control. The control set is generally placed on a support above the table, isolated from it. Inverted microscopes are most often utilized for patch-clamping cells or intracellular organelles, mainly because they allow the easy access of an electrode from the top. It is best to equip the microscope with a kind of contrast enhancement, such as Nomarski, Hoffman, or phase, each of which provides a sharp image of the contours of the vesicles and of the tip.

5.2 Electronics

As shown in *Figure 2*, the constituents of the electronic set-up are: an amplifier, a filter, an oscilloscope, a personal computer, an AD/DA converter, and a tape recorder. They are placed on a rack support, with the exception of the headstage (I-V converter) of the amplifier which stands inside the Faraday cage, directly connected to the pipette holder.

The amplifier (Axon Instr.; Bio-Logic; Dagan Co.; List-electronic) consists of two main parts, the I-V converter and the support circuits. Essentially, the former serves for imposing voltages and for measuring small currents with low noise by means of a high resistor R_f and the operational amplifiers A_1 and A_2 (see *Figure 3a*), while the latter further amplifies the current. Through the amplifier, one can either control the membrane potential and measure the evoked current (voltage-clamp mode), or impose a current and measure the membrane potential caused by the current (current-clamp mode). Experiments with mitochondrial membranes are usually carried out in the former mode. Through the knobs in the front panel of the amplifier's main unit it is possible to control the voltage (V_{hold}), the pipette offset (V_p-offset), the output scaling (G), and to compensate the pipette capacitive current (pipette capacitance compensation). In some computerized amplifiers, however, these parameters can be directly controlled by the computer.

The filter is used to remove high frequency noise from the recordings. In general, lowpass filters are used by which frequencies higher than those of the current signals are eliminated. Amplifiers can have a built-in filter. However, this can be a disadvantage because both current outputs are either filtered or not. On the contrary, if a separate filter (either Butterworth or Bessel) is available, one of the output currents from the amplifier can be filtered and observed on the monitors during the experiment, while the other can be recorded bypassing the filter. This system has the advantage that, at the stage of analysis, the records can be filtered in the most convenient way.

To visualize the electric activity of a membrane, one may use either an oscilloscope or a computer monitor. It is, however, advisable to have both, as the oscilloscope provides the best temporal resolution of the signal, but can not store data, while the opposite holds true for the computer. With the computer one can also apply complex experimental protocols as well as analyse data. Obviously, the computer can be used only if interfaced with an AD/DA converter (Instrutech Co.).

In general, the amount of data of a patch-clamp experiment under continuous mode would rapidly saturate the computer's disk capacity. For this reason, it is best to store the data using a tape recorder like, for example, a standard video tape recorder connected to the amplifier through the proper interface (PCM: pulse code modulator) (Instrutech Co.).

During an experiment, connections are as follows. One of the current outputs of the amplifier is connected to the PCM input channel (CH2), while the other goes to the filter and then to both the oscilloscope (CH2) and the computer (through the AD converter). The voltage output of the amplifier goes to the other channel of the PCM (CH1), to the AD converter and to the (CH1) channel of the oscilloscope. If the computer is the external stimulus source, connect one channel of the DA converter to the amplifier stimulation input. For replaying data using the video tape recorder (VTR), the PCM current output (CH2) is filtered, and then is again channelled to both the oscilloscope (CH2) and the computer (through the AD converter). The output signal of the stored voltage can go to the oscilloscope (CH1) and/or to the computer (through the AD converter).

By means of wires, it is necessary to ground all mechanical and electronic elements, as well as any conductive material. The wires are then grouped together at a single point, on the Faraday cage for example, which in turn is connected to the signal ground of the amplifier. The signal ground is linked to the power ground provided by the wall socket. Being a source of noise, it is recommended to isolate adequately the computer's monitor and to provide the computer with its own power line and ground.

6. Preparation of the patch-clamp electrodes and of the dish

The bridge connecting a membrane under analysis and the amplifier consists of a metal wire inserted into a glass envelope (pipette) filled with an electrolyte solution. Pipettes have to be prepared in the laboratory using suitable instruments, and also the commercially available silver wires need to be treated further and maintained with care.

6.1 Preparation of pipettes

Pipettes are prepared from glass capillaries. There are a wide variety of capillaries commercially available (the most commonly used have an internal diameter of 0.75–1.2 mm and an external diameter of 1.0–2.0 mm) made from glass with specific properties, which in turn will influence the choice of the capillary to be used for a particular experiment (single channel- or whole cell-recording). It is important to note that:

(a) Soft (soda) glasses are easier to forge and consequently it is easier to obtain tips with shapes suitable for good seals.

(b) Hard (borosilicate or aluminiumsilicate) glasses have electric properties (such as a low dielectric constant and high resistivity) which minimize electric noise and the capacitance of the electrode, and consequently improve the quality of the recording.

(c) A glass wall thicker than 0.3 mm provides a blunt tip and also lowers the noise.

(d) Glasses (soda, for example) may release compounds interfering with channel activity in different ways.

Once the type of glass has been chosen (for mitochondrial single channel recording pipettes, we recommend KG-33 (Kimax from Kimble Div.) or o.d. 1.7, i.d. 1.0 (Hilgenberg) glasses, composed of borosilicate with an intermediate softening temperature), suitable pipettes are prepared according to the following sequence (Sections 6.1.1 to 6.1.6).

6.1.1 Major equipment required

- Glass capillaries of around 7 cm long
- A puller
- Material of low dielectric constant
- ×30–50 magnification microscope and a platinum wire connected to a power supply
- A fire polisher

6.1.2 Preparation of glass capillaries

See *Protocol 8*.

Protocol 8. Preparation and cleaning of glass capillaries

1. Pass the two ends of a capillary (approx. 7 cm long) over a flame to round the rims. (This will prevent scratching the metal wire.)

2. Place the glass in a container with 100% ethanol, or 70% methanol, and mix thoroughly to ensure complete cleaning. If particularly dirty, sonicate with an ultrasonic cleaner. Subsequently great care must be taken to avoid touching with fingers the middle part of the glass, where the tips will be formed.

3. Decant the ethanol and let the glass dry in a hot oven for a few hours.

4. Store the glass in a closed container to avoid any contamination.

6.1.3 Pulling of glass capillaries

This operation allows one to obtain two pipettes from a single piece of glass, each of which has a blunt tip smaller than the original. It is achieved using an instrument called a puller (List-electronic; Narishige) and the procedure consists of at least two stages, in which pipettes are formed by combining the action of heat with that of a pulling force. In the first stage, the diameter of the capillary's centre is reduced and stretched; in the later stages the glass is broken yielding two tips. Microprocessor-driven pullers which can store multiple programs are also available (Sutter Instr.). It is difficult to write a general protocol because this will depend upon the puller available and the type of glass used. However, by following carefully the instructions provided by the different firms, then after some practise, pipettes with the desired properties can be obtained quite easily.

6.1.4 Coating of the tip

The noise current of single channel recordings (originating from the pipette connected to the headstage) can be substantially lowered by applying a hydrophobic material close to the pipette's tip (*Protocol 9*). Silicon polymers are most often used (as Sylgard 184 from Dow Corning or RTV 615 from General Electric), but enamel can apparently act as a good substitute. In any case, the electric noise arising from the tip diminishes because solutions are no longer drawn up the glass wall and also because the low dielectric constant of the material improves the electric properties of the painted glass.

The coating is done under a microscope at ×30–50 magnification. It is preferable to use optical fibres and to clamp the pulled pipette on to a separate support. If the illumination is properly orientated, the glass wall will stand out as bright lines in the dark field.

Protocol 9. Coating the pipette's tip with Sylgard

A. *Coating of the tip of the pipette*

1. Dip a metal hook, or a glass capillary forged into a hook, into the unpolymerized Sylgard warmed to room temperature.

2. Paint the zone near the tip, until approximately 0.1–0.5 mm from its end, by directing the Sylgard away from the tip. Simultaneously, rotate the pipette. Never coat the extreme part of the tip (which will be in contact with the membrane), otherwise the polishing step or the seal with the membrane will be unsuccessful.

3. Allow polymerization to occur by exposing the tip for 10–15 sec to a hot air stream opposite to it, or by placing the tip inside a slightly heated platinum coil. In either case rotate the pipette again in order to ensure uniform and complete polymerization. At the end of this process the Sylgard should be hard to the touch.

B. *How to prepare and store Sylgard*

1. Weigh ten parts of the silicone rubber and one part of the catalysing agent directly into a container, and mix thoroughly with a glass rod.

---

2. Place the mixture in an oven at 50–60°C to harden. The mixture should become more viscous without polymerizing; hence approximately every 2 min check the consistency of the material. Usually 8–12 min will suffice.

3. Store the Sylgard at − 10 to −20°C in closed containers as 0.1–0.3 ml aliquots.

4. For the coating step, bring it to room temperature and use it within a reasonable period of time before it polymerizes (approximately half an hour).

6.1.5 Fire polishing of the tip

The final stage of pipette preparation consists of the fire polishing of the tip in order to:

- completely remove impurities that can hamper the formation of good seals
- round the rim adequately to prevent any damage to the membrane
- regulate precisely the tip diameter

The apparatus for fire polishing (Narishige) consists of an upright microscope with a final magnification of ×300–500 and with long working distance objectives; a bent platinum wire connected to a current or voltage supply; a set of micromanipulators connected to either the pipette or the wire. It is also possible to buy more sophisticated equipment (List-electronic) which utilize inverted microscopes, where both the coating and polishing can be carried out. By direct observation of both the wire and the tip, the two are brought to within a few micrometres from each other. (Remember though that the metal expands upon heating!) As the wire is heated, the tip slightly melts thus reducing its internal diameter.

As for the pulling, precise protocols to obtain the desired tip dimensions have to be adjusted to the particular needs. However it is useful to remember to:

(a) Prevent deleterious contamination of the pipette by platinum sputtered from the wire, by always coating a new wire with glass. Allow a pipette to touch the overheated wire: the melted glass will form the desired covering.

(b) Carefully store fire polished pipettes in closed containers to avoid any contamination.

(c) Fire polish the pipettes just before use, even if sometimes it may be convenient to carry out the pulling and coating the day before.

6.1.6 Filling of the pipette with solutions

The filtered solution to fill pipettes for patch-clamp experiments of mitochondrial membranes is usually as follows: 150 mM KCl, 20 mM Hepes (pH 7.2 with KOH), 0.1 mM $CaCl_2$. The concentration of calcium ions is clearly unphysiological. However, if it has been established that the calcium ions do not affect the behaviour of the channel, the ions should be present as they increase the probability of forming tight seals.

A tricky step is the filling of the pipette's extremely thin tip, which can be accomplished by letting the solution enter by capillary action (see *Protocol 10*).

Protocol 10. Filling of the pipette with solutions

1. Clamp the pipette to a support with a clip. Cover the whole set-up, with an upside-down beaker for example, to avoid contamination by dust.
2. Immerse the tip perpendicularly in the filtered solution contained in a small beaker for 3–4 min. Be careful not to dip the pipette higher than the Sylgard-coated area: wet glass can act as a source of noise.
3. Release the pipette and using a syringe with a thin plastic tube (see section 6.3) backfill the remainder up to approximately half of the total volume, enough to have the electrode immersed (see section 6.2 and *Protocol 13*), but avoiding possible damage to the holder's interior through contact with the solution during mounting of the pipette.
4. Ensure under the microscope that no bubble is present in the tip. When present, hold the pipette with the thumb and the forefinger and flick the region close to the tip firmly with the other hand.

6.2 Preparation and maintenance of the electrodes

6.2.1 Measuring electrode

The measuring electrode consists of a silver filament coated with AgCl (Ag/AgCl) or of a commercially available Ag/AgCl pellet. It is inserted into the holder at one end, while the other dips into the solution of the pipette. If the internal pipette diameter can hold the pellet, it is convenient to use it because this removes the need for periodical chloruration. On the contrary, the silver wire must be prepared in the following way:

(a) Clear the filament of any oxide or other impurities using a fine sandpaper, then rinse with 70% ethanol or methanol.

(b) Keep the filament fully immersed in Chlorox bleach for 20–30 min to produce the desired AgCl (grey-black) coating, then rinse with bidistilled water.

The filament is now a good electrode. However, with time, and also due to damage caused by mounting the pipette into the holder, the coating can be disrupted. Whenever this happens, the coating operation must be repeated.

The holder, made of inert, low dielectric constant material, is the part of the patch-clamp set-up which physically links the electrode to the amplifier through a gold-plated pin. It also has a well where the glass pipette is accommodated. It should be kept perfectly clean to allow low noise recordings. Hence, from time to time:

- dismount the holder and soak it in methanol or ethanol for a few minutes
- dry it with pressured air or nitrogen, better if filtered

6.2.2 Ground electrode

Similar to the measuring electrode, the ground electrode can be Ag/AgCl filament or pellet. It connects the bath solution to the ground and therefore it dips into the solution at one end, while the other is linked to the headstage through a plug. If chloride solutions are used, then during the course of an experiment the chloride activity in the immediate vicinity of the ground electrode may not remain constant, or the temperature of the bath may change. The appearance of a voltage offset between the two electrodes can thus be prevented by placing the ground electrode a safe distance away from the bath, i.e. at the end of a 3–6 cm long KCl solution–agar bridge (*Protocol 11*).

Protocol 11. Preparation of the agar bridge

1. Prepare 25 ml of 1 M KCl (or of the KCl concentration generally used in the bath).
2. Dissolve 2% by weight of agar–agar into the above solution by stirring with gentle heating.
3. Bend one end of a plastic tube (an insulin syringe without the plunger and truncated to the desired length, for example) over a flame.
4. Place the bent end in the KCl–agar mixture, and suck up the mixture using a syringe connected to the other end.[a]
5. Insert the electrode in the bridge whilst it is still liquid and let it solidify.

[a] This reduces the formation of unwanted air bubbles.

6.2.3 Maintenance of the electrodes

After use, the electrodes are stored dipping into the KCl solution used for preparing the agar bridge. It may be a good practice to connect the two electrodes in order to keep them equilibrated.

6.3 Preparation of syringes for filtering and adding solutions

Equip plastic syringes with 0.22 μm filter units in order to remove any contaminants from the solutions. The needle is prepared from an elongated yellow tip as follows:

(a) Hold the tip at the two ends and place the middle part over a flame. As it starts to melt, remove the tip from the flame and simultaneously pull the two extremes so that the central part stretches into a thin tube.

(b) Cut the filament as desired and insert the wider end into the filter unit.

6.4 Preparation of the dish

See *Protocol 12.*

Protocol 12. Preparation of the dish for the patch-clamp experiment

1. Use plastic Petri dishes of 35 mm diameter. It is best to replace the centres with glass (for example coverslips) in order to increase mitochondrial adhesion to the bottom and to reduce the focal distance. To do this:

   (a) Paint with Sylgard (prepared as in *Protocol 9B*) the external rim of a 10 mm hole cut in the middle of the dish.

   (b) Press the coverslip on to the rim.

   (c) Let the dishes stand upside-down in an oven at 50–60°C overnight or until the adhesion is complete.

   To ensure that the dishes are perfectly clean:

   (a) Wash the dishes after use under tap-water and then leave them to soak.

   (b) Use diluted liquid soap and remove any debris with a cotton Q-tip to clean them further.

   (c) Rinse well with bidistilled water and then with ethanol.

   (d) Dry the dishes under a stream of air and keep them in a closed container.

2. Place 5–10 μl of the experimental suspension in the centre of the dish, and inject the experimental medium starting from the border of the dish until the liquid converges naturally over the membrane suspension. In this way, vesicles remain more concentrated in the centre. 1 ml is routinely used, but if a smaller volume is preferred (0.2–0.5 ml), inject the medium starting from the outer rim of the suspension drop.[a]

3. Place the dish under the inverted microscope of the patch-clamp set-up and look at the mitochondria or mitoplasts with ×600 magnification or higher. If proteoliposomes are on the dish, ×500 (or even less) magnification will suffice.

---

4. Wait approximately 5 min then check that vesicles have properly sedimented. Perfuse the dish extensively either with a perfusion set-up or, manually, by withdrawing and injecting simultaneously the experimental medium with two syringes placed at opposing ends of the dish. Well adhered vesicles should not move.

5. Choose a vesicle to patch. Any type of vesicle should adhere perfectly to the glass and have a round shape with a well contrasted membrane.[b] Suitable mitoplasts are easy to spot and select because of the frequent presence of a clear, dark cap region while mitochondria look smaller and dark all over (see *Figure 1*). Among liposomes, whether multi- or unilamellar, those with intermediate size (approx. 4–7 μm of diameter) are usually easier to patch.

---

[a] For patching mitochondrial membranes under symmetrical conditions, the bath solution is generally that described in section 6.1.6.

[b] Be aware that sometimes it is difficult to make seals in spite of the appearance of the vesicles, i.e. high temperature, or other factors, can produce sticky membranes.

7. Patching the membrane

See *Protocol 13*.

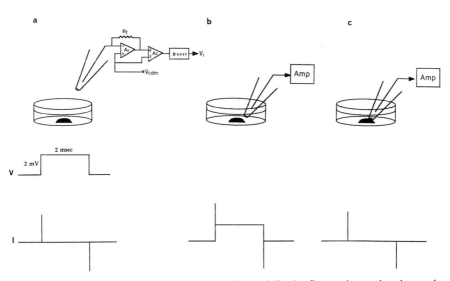

Figure 3. Sequential steps to obtain a patch. Essentially the figure shows the change in current (*I*) (lower panel) in response to the voltage pulse (*V*) imposed to a patch-clamp electrode (upper panel) not yet dipped in the bath solution (a), or immersed (b), or after seal formation (c). For details, see *Protocol 13*. In the upper panel of (a), the major constituents of the amplifier (Amp) I-V converter are shown (see section 5.2).

Protocol 13. Operational protocol for patching the membrane

1. Place the dish on the microscope plate and select an organelle or liposome suitable for patching (see *Protocol 12*).

2. Immerse the ground electrode in the dish.

3. Insert the pipette into the holder, adequately tilted (see section 5.1).

   (a) Pay attention not to scratch the filament.

   (b) Ensure that the filament dips into the solution.

   (c) Fix the pipette securely.

   (d) Impose a positive pressure, by blowing through the tube by mouth, in order to prevent floating membranes or dirt sticking to the tip. Keep the pressure on, by acting on the switch (see *Figure 2*).

4. Enter the computer acquisition program and impose a voltage pulse (usually of 2 mV for 2 msec) (or use the seal test given by the amplifier). At this stage no current is observed (see *Figure 3a*).

5. Using the macromanipulators, drive the tip of the pipette as close as possible to the centre of the light spot on the dish (it may help to have the hole of the lens below the dish as a reference) and dip it into the solution. At the precise moment of immersion, the current flows through the electrode as clearly observed on the oscilloscope and/or on the computer video[a] (see *Figure 3b*).

6. Calculate the resistance of the pipette by applying Ohm's law. Hence divide the magnitude of the imposed pulse by the observed current (see step 5), accurately measured with the proper gain (as 5–10 mV/ pA, for example) and scale, and with the filter set at 5–10 kHz.[b]

7. Bring the tip into focus. Some practice is needed to accomplish this step. Amongst the possible strategies, one way is to:

   (a) Bring the focus up slowly (from the plane of the vesicles) and look for the pipette's shadow by moving the pipette using the macro-manipulator.

   (b) Manoeuvre the focus and the pipette, find the tip, and place it on focus.

8. Reach the vesicle by:

   (a) Setting the focus slightly below the tip.

   (b) Bringing the tip again into focus using the macromanipulators.

   (c) Repeating this operation with care until the tip is just slightly above the plane of focus of the vesicles (4–5 μm).

9. Release the pressure. However, if the pressure is not too strong and/ or if the vesicle adheres firmly to the bottom of the dish, it may be left on until the pipette is almost in contact with the membrane.

10. Use the micromanipulators for the final stages of the pipette's movements in order to position the tip properly with respect to the vesicle, i.e. in the upper right quadrant of the vesicle (see *Figure 3c*). Remember that a seal is more difficult to form if the tip has previously touched the membrane.

11. In spite of the presence of symmetrical solutions, a small offset potential frequently exists between the two electrodes (visualized when the zero current signal does not lie on the zero line). Hence, when close to the membrane, increase the amplifier gain (at 10 mV/pA, for example) and zero the pipette current using the pipette offset knob of the amplifier.

12. Touch the membrane and observe a slight decrease of the resistive current. Though a very tight (gigaohm) seal can form spontaneously, it is frequently necessary to:

    (a) Apply a negative pressure (suction by mouth) to achieve the so-called Ω-shaped patch.

    (b) Apply a small negative potential (−10 mV, −20 mV) to increase the probability of forming successful seals.

13. Whilst the seal forms, observe a continuous decrease in the resistive current until it almost completely disappears (see *Figure 3c*).

14. Compensate the pipette capacitive current still evident (see *Figure 3*) (mainly due to the electric properties of the pipette wall and of the membrane) with the fast and slow time constant compensators of the amplifier.

15. Switch off the microscope's light and close the Faraday cage with the screen. Set the filter to 1–2 kHz, and increase the sensitivity of the scale. The patch is now ready to be studied in the cell attached configuration.[c]

16. Set the amplifier in the voltage-clamp mode (see section 5.2), and then switch the computer to the continuous mode in order to detect channel activity (if present) under steady state conditions at the voltage imposed through the computer. Depending upon the behaviour of the membrane, you may apply also voltage pulses or ramps utilizing the appropriate programs.

17. Use a video tape recorder to store real time current data. Only two signals can be stored. You may use the second input to record the imposed voltages; then have someone to write down the timing of the distinct stages of an experiment so that, when data is replayed for analysis, you follow exactly on the video tape the time at which an event has occurred (a voltage change, or an addition, etc.). Alternatively, record the voice through the second input.

Protocol 13. *Continued*

18. If you want to have an excised inside out patch.[d]

   (a) Withdraw the pipette from the vesicle with a firm upwards movement by means of the micromanipulator. If, either the entire vesicle moves with the pipette, or it breaks into smaller vesicles, then:

   (b) Disrupt the vesicle by driving the pipette out of the bath solution and then re-immersing it quickly.

   In any case, confirm the integrity of the seal by looking at the seal test.

19. Make the desired additions (after setting the gain to the lowest value):

   • by adding microlitre volumes close to the patch using an automatic pipette

   • by perfusing at least three to five times the bath's volume with the perfusion set-up

[a] Remember that failure to detect current at this stage can be due to:
   • failure to immerse the ground electrode in the bath solution
   • a bubble in the pipette tip
   • a clogged tip
   • a faulty connection between the holder and the amplifier

[b] As there is a close relation between the resistance of a pipette and the tip opening diameter:
   • resistances for mitochondrial single channel experiments commonly range from 10–20 MΩ (in approximately 150 mM salt)
   • resistances for whole mitoplasts[c] recordings should have less than 10 MΩ

[c] The cell attached configuration of the patch is the first to be obtained and is the starting configuration to produce the other types of patches (2). Excision of the membrane patch under the pipette yields the inside out configuration, which has the advantage of exposing the internal side of the membrane to the bath solution where additions can be easily made. Conversely, disruption of the patch under the pipette yields the whole cell configuration, from which an outside out patch can be obtained. With mitoplasts or mitochondria, inside out patches are quite easy to obtain. On the contrary, application of the usual tricks employed to obtain the whole cell configuration (i.e. imposition of high voltage pulses in the millisecond range) are largely unsuccessful. Fortuitously, however, whole mitoplast patches do form spontaneously, as detected by:
   • an increase in the total current at a given potential
   • an increase in the membrane's capacitance together with a decrease in resistance

To the best of our knowledge, outside out patches of mitochondrial membranes have never been achieved. However, as there is no theoretical obstacle specific to mitochondrial membranes, one may wish to follow the protocols reported in ref. 2.

[d] With liposomes, especially if multilamellar, it is highly advisable to work with this configuration.

Acknowledgements

This work was supported by grants of the Consiglio Nazionale delle Ricerche and of the Ministero dell' Università e della Ricerca Scientifica e Tecnologica of Italy.

References

1. Sorgato, M. C. and Moran, O. (1993). *CRC Crit. Rev. Biochem. Mol. Biol.*, **18**, 127.
2. Sakmann, B. and Neher, E. (ed.) (1985). *Single-channel recording.* Plenum Press, New York.
3. Rudy, B. and Iverson, L. E. (ed.) (1992). *Methods in enzymology*, Vol. 207. Academic Press, Inc.
4. Hille, B. (1992). *Ionic channels of excitable membranes.* Sinauer Associates Inc., Sunderland, Massachusetts.
5. Suzuki, K. (1969). *Science*, **163**, 81.
6. Ausubel, F. M., Brent, R., Kingston, R. E., Moore, D. D., Seidman, J. G., Smith, J. A., and Struhl, K. (ed.) (1987). *Current protocols in molecular biology*, Vol. 2, Chapter 13. Greene Publishing Associates, Inc. and John Wiley and Sons, Inc.
7. Daum, G., Böhni, P. C., and Schatz, G. (1982). *J. Biol. Chem.*, **257**, 13028.
8. Criado, M. and Keller, B. U. (1987). *FEBS Lett.*, **224**, 172.
9. Labarca, P., Coronado, R., and Miller, B. C. (1980). *J. Gen. Physiol.*, **76**, 397.
10. Kagawa, Y. and Racker, E. (1971). *J. Biol. Chem.*, **246**, 5477.
11. Sandri, G., Siagri, M., and Panfili, E. (1988). *Cell Calcium*, **9**, 908.

8

Studies of cellular energetics using ^{31}P NMR

KEVIN M. BRINDLE, ALEXANDRA M. FULTON, and
SIMON-PETER WILLIAMS

1. Introduction

Nuclear magnetic resonance (NMR) has become an established technique for the non-invasive study of metabolism in systems ranging from microbial cells to the human brain. It is a tremendously versatile technique in terms of the nuclei that can be monitored and thus the type of problems that can be addressed. For example flux of material in a metabolic pathway can be monitored by feeding the system a ^{13}C-labelled substrate and then monitoring the subsequent isotope redistribution using ^{13}C NMR (1). Ion fluxes and gradients can be monitored using ^{23}Na and ^{31}K NMR or by using ^{87}Rb NMR in studies where Rb^+ is used as a replacement for K^+ (2). The intracellular concentrations of cations such as Ca^{2+} and Mg^{2+} can be monitored using ^{19}F NMR in tissues which have been loaded with fluorine labelled chelating agents which bind these ions (3). ^{1}H NMR can be used to monitor the levels of a variety of cellular metabolites and, in imaging studies, to determine the spatial distribution of water and hence the structure of the system under study. Although all these nuclei can be used to a greater or lesser extent to study cellular bioenergetics we will restrict our discussion here to the nucleus which has made the greatest contribution in this area, the naturally abundant isotope of phosphorus, ^{31}P.

The observation that the chemical shift of the inorganic phosphate (P_i) resonance in the ^{31}P NMR spectrum of erythrocytes could be used to determine the intracellular pH (4), followed by the realization that these non-invasive measurements could also be made on skeletal muscle (5) led, in the early 1970s, to the birth of 'in vivo NMR'. Other major landmarks in this field were the introduction of spatial localization techniques, with the advent of the surface coil in 1980 (6), and horizontal large bore magnets, which facilitated measurements on animals and allowed the first studies on human limbs (7). These were followed by very large bore magnets which could accommodate the whole human body. Subsequent developments in localized spectroscopy mean that it is now possible, in humans, to acquire ^{31}P spectra from specific regions of organs such as the brain. Similar and parallel developments occurred in NMR studies of other nuclei, particularly ^{13}C and [1]H.

2. Methodology

2.1 Basic principles of the NMR experiment

A full description of the NMR experiment is beyond the scope of this chapter. Instead we will confine the discussion to some of the more basic concepts, in particular those that are directly relevant to the ^{31}P NMR experiment *in vivo*. More detailed descriptions can be found in refs 8–10, amongst others.

Nuclei with non-zero nuclear spin possess a magnetic moment. In the presence of an applied magnetic field these nuclei assume a number of discrete energy levels, which are determined by their spin quantum number. For nuclei with a spin quantum number of ½, which includes ^{31}P and some of the other nuclei commonly studied *in vivo*, the spins have only two allowed energy levels in which the nuclear magnetic moments are aligned with or against the applied field, see *Figure 1*. These are the low and high energy levels respectively. The magnetic moments of the nuclei do not align exactly with the applied field but precess about it. The frequency of this precession,

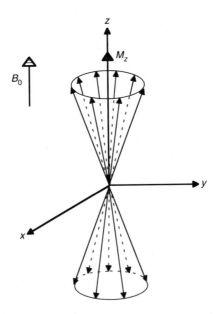

Figure 1. Orientation and precession of nuclear spins in a static magnetic field (B_0). In this representation the precessing spins in the sample have been brought to a common origin. The slight excess of spins aligned with the field leads to a net magnetization vector which lies along the positive z axis.

the Larmor frequency, is dependent on the gyromagnetic ratio of the nucleus (γ) and the strength of the applied field (B_0). The gyromagnetic ratio is a proportionality constant relating the precessional frequency, ν, (in hertz) of a particular nucleus to the applied field, i.e.

$$\nu = \frac{|\gamma|B_0}{2\pi}.$$ [1]

Thus at a field of 9.4 Tesla the precessional frequency of ^{1}H is 400 MHz, of ^{31}P is 162 MHz, of ^{13}C is 101 MHz, and of ^{15}N is 41 MHz. (Note the earth's magnetic field is approximately 60 μT.) The difference in energy (ΔE) between the two allowed energy levels is given by:

$$\Delta E = h\nu$$ [2]

where h is Planck's constant and ν is the Larmor frequency. At thermal equilibrium the energy levels are populated according to the Boltzmann equation:

$$N_{upper}/N_{lower} = \exp(-\Delta E/kT)$$ [3]

where N_{upper} and N_{lower} represent the populations of the nuclei in the upper and lower energy levels respectively, k is the Boltzmann constant, and T is the absolute temperature (in K). As the energy difference is very small the energy levels are nearly equally populated with a slight excess of spins in the lower energy level, i.e. aligned with the B_0 field. This excess of spins in the lower energy level means that in a large population of spins the net magnetization vector lies in the B_0 direction, or along the z axis. This is referred to as the z magnetization (M_z) and is a bulk property of the sample (see *Figure 1*). Application of an oscillating magnetic field (B_1) which is perpendicular to the static B_0 field, i.e. in the transverse or x, y plane and which has the same frequency as, or is resonant with, the Larmor frequency of the nucleus, induces transitions between the two energy levels. A net transfer of spins from the lower to the higher energy level results in absorption of energy and it is this which is detected in the NMR experiment. Another way of looking at this is in terms of the bulk magnetization (see *Figure 2*). The net magnetisation (**M**), which initially lies along the z axis, is tipped by the applied field (B_1) into the x, y plane (*Figure 2b*). It is while it is in this plane that it is detected by the NMR receiver. Thus if we tip the net magnetisation so that it lies totally in the x, y plane, the z component of the magnetization will have been destroyed and the spin populations of the two energy levels will have been equalized. Similarly if the net magnetisation is tipped until it comes to lie along the $-z$ axis the spin populations will have been inverted (*Figure 2c*).

At the field strengths commonly in use, the nuclei of different elements resonate at frequencies differing by several tens of megahertz. However the nucleus of any one element will resonate not at a single frequency but over a range of frequencies determined by its chemical environment. This is

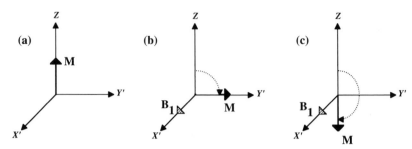

Figure 2. A rotating coordinate system can be defined in which the x' and y' axes are rotating about the z axis at the operating frequency of the spectrometer. Thus an oscillating magnetic field (B_1) applied at this frequency and perpendicular to the B_0 field will appear as a static vector in the x' y' plane. Application of the B_1 field along the x' axis causes the net magnetization vector (**M**) to rotate about it in the y, z plane. The angle of rotation will depend on the gyromagnetic ratio of the nucleus, the amplitude of the applied field, and its duration or pulse width. (b) The field has been applied for long enough to rotate the net vector, **M**, through 90° and is thus referred to as a $\pi/2$ or 90° pulse. Application of the B_1 field for twice as long results in inversion of the magnetization (c).

because the magnetic field experienced by each nucleus is modified slightly by the surrounding electrons. Differences in frequency due to this effect range up to a few kilohertz for most nuclei. Thus the phosphorus nuclei in the γ-phosphate of ATP resonate at a slightly different frequency from the phosphorus nuclei in phosphocreatine (see *Figure 3*). Since the resonance frequency depends linearly on the strength of the applied magnetic field and thus varies with the spectrometer used, it is usual to express the frequency of each peak or resonance in parts per million (p.p.m.) of the operating frequency of the spectrometer. This gives a value which is independent of the magnetic field strength, B_0, and is called the chemical shift (δ).

$$\delta = \frac{(\nu - \nu_{\text{ref}})}{\nu_{\text{ref}}} \times 10^6 \qquad [4]$$

where ν_{ref} is the operating frequency of the spectrometer and ν is the frequency of a resonance. The signals shown in *Figure 3* were acquired simultaneously. This is because the oscillating magnetic field was applied in the form of a pulse of a few microseconds duration. This had the effect of simultaneously exciting all the resonances within a few kilohertz either side of the frequency of the pulse. The limited bandwidth of the excitation pulse means, for example, that ^{1}H resonances could not be excited in a ^{31}P NMR experiment. An additional restriction is imposed by the limited bandwidth of the NMR excitation/receiver coil. However signals can be acquired concurrently from more than one element by using multiply tuned NMR probes and by employing appropriate radiofrequency and data routing in the spectrometer. The transient signal, termed the free induction decay (f.i.d.), obtained after

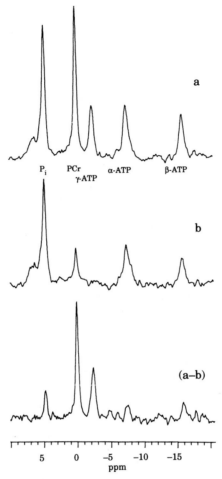

Figure 3. ^{31}P NMR magnetization transfer measurements of the P$_i$→ATP and phospho-creatine (PCr)→ATP fluxes in the hindlimb muscles of the rat. (a) A control spectrum. (b) The γ-phosphate resonance of ATP (γ-ATP) was selectively saturated leading to its loss from the spectrum. Transfer of this magnetic label from ATP to P$_i$ and phosphocreatine, via chemical exchange, results in a decrease in their resonance intensities. This is most clearly seen in the difference spectrum (a−b). These spectra were adapted, with per-mission, from ref. 13.

the excitation pulse, contains the frequency and amplitude information for the different resonances. This information can be extracted by Fourier trans-formation to produce a spectrum which has axes of amplitude versus frequency (see *Figure 3*). The intensity or area of a peak in this spectrum is proportional to the number of nuclei giving rise to it and thus the concentration of the metabolite containing the nucleus. However the relative intensities are also influenced by the spin–lattice relaxation times (T_1s) of the resonances. Following the radiofrequency pulse and excitation of the transient NMR signal, the spins relax back to their equilibrium populations, or the bulk magnetization vector returns fully along the z axis, with a time constant, T_1. If a resonance is excited again before equilibrium has been re-established then the signal can become saturated and its intensity reduced (see below). Relative intensities can also be affected if excitation by the radiofrequency pulse is not uniform across the spectrum. However for nuclei such as ^{31}P, which has a relatively narrow chemical shift dispersion, provided that the pulse is relatively short and of high power, this problem can be ignored.

Relaxation of the spins back to equilibrium can be characterized by two exponential time constants; T_1, the spin–lattice relaxation time and T_2, the spin–spin relaxation time. T_2 describes the lifetime of the transverse or x,y magnetization created after the excitation pulse. In the absence of B_0 inhomogeneity, the T_2 of a resonance determines its linewidth in the spectrum obtained after Fourier transformation, i.e.

$$\Delta \nu = \frac{1}{\pi T_2} \qquad [5]$$

where $\Delta \nu$ is the linewidth at half peak height. Both T_1 and T_2 can vary for nuclei in different chemical environments and will depend on a number of factors. For example as the rate of tumbling of the molecule decreases, i.e. as its correlation time increases, the T_2 of its resonances will get progressively shorter and as a result their linewidths will get larger. It is often assumed that binding of small molecules, such as ADP, to large molecules, such as proteins, leads to extreme broadening and thus loss of their resonances from the spectrum. However, in this context it is perhaps worth remembering that ^{31}P NMR signals have been detected from enzyme bound substrates (11).

Resonance linewidths may also be affected by exchange of the nucleus between two different chemical environments in which it has different resonant frequencies. When the exchange rate is fast compared to the difference in these frequencies then a single resonance is observed, the frequency of which is an average of these two frequencies and is weighted according to the relative populations of the nuclei in each site. This is the situation with inorganic phosphate whose resonance frequency is a weighted average of the resonance frequencies of the different protonated forms. If the exchange rate is slow compared to the difference in frequency then two separate resonances are observed. This is the case with some of the ^{19}F NMR probes of cell calcium, which show separate calcium bound and free resonances (3). At intermediate rates of exchange, however, the resonance is broadened. This can occur in ^{31}P NMR with Mg^{2+} binding to ATP and via enzyme catalysed exchange (11). Thus the T_2s or linewidths in NMR spectra are affected by the mobility of the nucleus, interaction with paramagnetic ions, and by chemical exchange.

T_1s are also affected by the factors which affect T_2. As T_1 is always longer than T_2 it limits the rate at which data can be acquired. In liquids the two may be comparable but in ^{31}P NMR experiments *in vivo*, T_2 is almost invariably much shorter than T_1. If a pulse is applied to a sample at thermal equilibrium, which rotates all of the z magnetization into the transverse plane, a so-called 90° or $\pi/2$ pulse, then the maximum amount of signal will be elicited (see *Figure 2*). However if another pulse is applied before equilibrium is re-established, the z magnetization (M_z) will be smaller and the amount of signal obtained correspondingly reduced, i.e. the signal will be partially saturated. The steady state z magnetization obtained following a series of pulses is given by:

$$M_z = M_0 \frac{(1 - e^{-T_r/T_1})}{(1 - e^{-T_r/T_1} \cos \alpha)} \qquad [6]$$

Where M_0 is the z magnetization at thermal equilibrium, α is the pulse flip angle, and T_r is the pulse repetition rate. For $\pi/2$ pulses this simplifies to:

$$M_z = M_0 (1 - e^{-T_r/T_1}). \qquad [7]$$

This shows that if M_z is to attain its equilibrium value between pulses, and therefore for the relative peak intensities in the system to be directly proportional to the concentration of the nuclei giving rise to them, then T_r must be set to four to five times the longest T_1 in the sample. This means in practice, with ^{31}P T_1s *in vivo* in the region of a second, data must be acquired quite slowly. If on the other hand improving the signal-to-noise ratio is more important than quantitation, then the maximum amount of signal should be acquired in as short a time as possible. This can be achieved by optimizing the pulse flip angle (α) for a given repetition rate. This optimum angle, known as the Ernst angle, is given by:

$$\cos \alpha = e^{-T_r/T_1} \qquad [8]$$

where the repetition rate, T_r, is equal to the acquisition time, which is the time taken to acquire the transient NMR signal following the pulse. This equation applies provided that $T_2 < T_r$. As ^{31}P T_2s *in vivo* are quite short, this is not a significant constraint.

We have now covered the very basic concepts of NMR as they relate to the practicalities of acquiring simple, non-localized, ^{31}P NMR spectra from living tissues. However for anyone seriously contemplating studies of this sort we recommend that you consult an experienced spectroscopist. We will now briefly describe a technique, unique to NMR, which can be used, *in vivo*, to measure exchanges catalysed by enzymes and membrane transporters.

2.2 Magnetization transfer measurements of exchange

A magnetization transfer experiment is analogous in many respects to an isotope exchange experiment in which the label is introduced non-invasively

Kevin M. Brindle et al.

in the form of a transient change in nuclear polarization. Consider a two-site exchange system in which a nucleus exchanges between two sites, A and B, in which it has different chemical shifts. For example, A and B might be the phosphorus nuclei in phosphocreatine and the γ-phosphate of ATP, which are in exchange in the reaction catalysed by creatine kinase.

$$A \underset{k_{-1}}{\overset{k_{+1}}{\rightleftharpoons}} B$$

The z magnetization of the nucleus in site A (M_A) and site B (M_B) can be described by coupled differential equations of the form:

$$dM_A/dt = -1/T_{1A} (M_A - M_0^A) - k_{+1}M_A + k_{-1}M_B \quad [9]$$

$$dM_B/dt = -1/T_{1B} (M_B - M_0^B) - k_{-1}M_B + k_{+1}M_A. \quad [10]$$

The last two terms of these equations simply describe chemical exchange between A and B. The first terms describe the rate at which the z magnetizations or spin populations of the energy levels of the A and B spins regain equilibrium via spin–lattice relaxation following a perturbation.

There are several NMR experiments which can be used to measure the exchange rate constants, k_{+1} and k_{-1} in this system (reviewed in ref. 12). In the interests of brevity we will confine our discussion to an experiment which is often used in ^{31}P NMR studies *in vivo*, the so-called 'saturation transfer' experiment. If the nucleus in site A is selectively excited, such that at $t = 0$, the z magnetization is saturated, i.e. $M_A = 0$ and the spin populations are equalized, then over a period of time the resonance intensity of the nucleus in site B will decay exponentially with a time constant $1/T_{1B} + k_{-1}$ and reach a new equilibrium value, M_∞^B, where:

$$M_\infty^B = \frac{1/T_{1B} \, M_0^B}{(1/T_{1B} + k_{-1})}. \quad [11]$$

Selective saturation of A is obtained by applying a relatively low power pulse at As resonant frequency. By acquiring spectra in which the saturation pulse is applied to A for various periods of time and fitting the intensities of B to a decaying exponential function, the sum $1/T_{1B} + k_{-1}$ can be determined (a variant of this experiment is described in *Protocol 4*). Comparison of the equilibrium value of B (M_∞^B) with the equilibrium or fully relaxed value obtained in the absence of saturation of A (M_0^B) allows calculation of T_{1B} and thus the first order exchange rate constant, k_{-1}. Multiplication of this by the concentration of B gives the flux from B to A. Equation 11 shows that in order for the exchange to be measurable, k_{-1} must be comparable with $1/T_{1B}$, the rate at which the magnetic label decays through spin–lattice relaxation. In ^{31}P NMR, T_1s are around a second and therefore this limits the technique to measurement of relatively fast exchanges in which the first order rate constants are of the order of 1/sec. Application of this technique to

166

measurement of the creatine kinase catalysed flux between phosphocreatine and ATP in the hindlimb muscles of the rat is illustrated in *Figure 3*. Saturation of the γ-phosphate resonance of ATP results in its loss from the ^{31}P NMR spectrum (*Figure 3b*). Transfer of this magnetic label to phosphocreatine results in a decrease in its resonance intensity (compare *Figures 3a* and *b*), the time course and magnitude of which can be used to estimate the flux from phosphocreatine to ATP (see above). The spectrum shown in *Figure 3b* is the steady state spectrum obtained following prolonged saturation of the γ-phosphate resonance.

3. ^{31}P NMR experiments *in vivo*

3.1 The type of information obtainable from ^{31}P NMR experiments *in vivo*

^{31}P NMR is ideally suited to the study of cellular bioenergetics. This is illustrated by the spectra of the hindlimb muscles of the rat shown in *Figure 3*. The chemical shift or frequency of the P_i resonance relative to some standard can be used, by comparison with an appropriate titration curve constructed from measurements *in vitro*, to estimate the intracellular pH. In muscle the standard often used is the phosphocreatine resonance, whose chemical shift does not titrate with pH in the physiological range (14). When the hindlimb muscles are stimulated to contract the P_i resonance shifts upfield, indicating acidification (*Figure 3*). The resonance is also very broad indicating pH heterogeneity within the muscle. In some systems P_i resonances from cellular compartments with different pHs can be distinguished. For example in yeast, vacuolar and cytoplasmic P_i signals can be observed (see *Figure 4*) and in heart muscle there is evidence of a separate mitochondrial P_i resonance, indicative of an alkaline pH in the mitochondrial matrix (15).

The concentration of ATP in the muscle can be estimated from the intensity (or area) of the β-phosphate resonance. There is a contribution to this signal from other nucleoside triphosphates such as GTP, but in most systems, including muscle, these are relatively low in concentration. The γ- and α-phosphate resonances of ATP are overlapped by the β- and α-phosphate resonances of ADP respectively and in studies on some tissues, such as liver, the ADP concentration has been estimated from the difference in intensity of the observed γ- and β-phosphate resonances of ATP. However, despite the fact that there are significant amounts of extractable ADP in most tissues the ADP resonances are usually very low in intensity. Thus this measurement must be considered as unreliable, relying as it does on measuring the small difference in intensity of two large ATP signals. The low ADP signal intensities in muscle are thought to be due to binding of ADP to actin and, in tissues like the liver, to binding to cellular enzymes present in high concentration and to sequestration within mitochondria. The degree of NMR visibility of

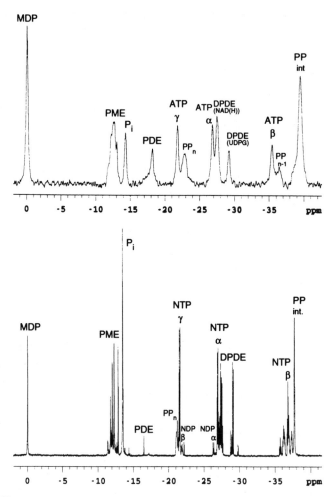

Figure 4. ^{31}P NMR spectra of intact, perifused, yeast cells (*top*) and a perchloric acid extract of the same cells (*bottom*). Abbreviations; MDP, methylenediphosphonate: γ-, α-, and β-NTP, the γ-, α-, and β-phosphate resonances of nucleoside triphosphates respectively; PME, phosphomonoesters, predominantly the sugar phosphates glucose-6-phosphate and fructose 1,6-bisphosphate, and nucleoside monophosphates; PDE, phosphodiesters; PP, polyphosphate where PP_n is from the terminal phosphates, PP_{n-1}, from the penultimate phosphates, and PP_{int} from the internal phosphates of long chain polyphosphates; DPDE, diphosphodiesters, including NAD^+ and NADH, and UDP-linked sugars, e.g. UDP-glucose.

ATP and ADP within mitochondria, however, is an unresolved issue. Although there is evidence in the perfused liver of 'NMR invisible' mitochondrial ATP (16), this is disputed by a more recent study (17). Furthermore these nucleotides are readily detectable in isolated mitochondria (18).

During the initial stages of muscle contraction, ATP levels are maintained at the expense of phosphocreatine (PCr) hydrolysis in the reaction catalysed by creatine kinase.

$$H^+ + PCr^{2-} + MgADP- \Leftrightarrow MgATP^{2-} + creatine \qquad [12]$$

If the muscle is stimulated to contract repeatedly at a level which, in the short-term, does not lead to fatigue (as in the hindlimb muscles used in the experiment illustrated in *Figure 3*), then a steady state will be achieved in which the ATP level is unchanged compared to the resting state and the phosphocreatine concentration is reduced. The level of phosphocreatine will be determined by the steady state ADP/ATP ratio. The phosphate lost from phosphocreatine appears as P_i. On cessation of contraction, the phospho-creatine is re-synthesized, the P_i level drops, and the pH returns to normal (approximately pH 7.05). The rate of PCr re-synthesis to its resting level has been used to monitor mitochondrial function in muscle (2, 7, 14). Flux in the creatine kinase reaction at rest or during steady state contraction can be monitored directly *in vivo* using magnetization transfer techniques, as described above and illustrated in *Figure 3*. The ease with which this measurement can be made has led to numerous studies of creatine kinase kinetics *in vivo*. The contribution that these have made to understanding the function of creatine kinase is reviewed in ref. 19. These studies have shown that the enzyme, which is located predominantly in the cell cytoplasm, is, under most con-ditions, catalysing a reaction which is near-to-equilibrium. This means that the free cytosolic ADP concentration, which is too low in concentration to be directly detectable, can be determined from ^{31}P NMR measurements of the enzyme's other substrates, i.e.

$$[\Sigma ADP] = [\Sigma ATP] \, [creatine]/K_{obs} \, [phosphocreatine]. \qquad [13]$$

The total concentrations or the sum of the different protonated and Mg^{2+} bound forms of ATP and phosphocreatine can be determined from the intensities of their resonances in the ^{31}P NMR spectrum. The creatine concen-tration can be calculated by subtracting the phosphocreatine concentration measured by ^{31}P NMR *in vivo* from the total creatine pool (creatine plus phosphocreatine) measured in tissue extracts. The equilibrium constant of the reaction, K_{obs}, is dependent on the pH and free Mg^{2+} concentration and can easily be calculated if these are known (20). The free Mg^{2+} concentration can be determined from the chemical shifts of the γ- and β-phosphate reso-nances of ATP, as these shift with Mg^{2+} binding to the nucleotide. However these shifts in the ATP resonances are relatively small over the physiological range of Mg^{2+} concentrations and this can compromise the precision of the measurement. This estimation of the free ADP concentration can be made in any tissue which expresses creatine kinase; this includes muscle and neural tissue and some tumours. The measurement can also be made in tissues which do not normally express the enzyme by introducing it using molecular genetic techniques. Expression of the enzyme in livers of transgenic mice showed that the free ADP concentration was much lower than the total extractable concentration (21).

The free cytosolic ADP concentration is an important parameter to be able to determine, in view of its role in the control of cellular energetics. Together with measurements of the ATP and P_i concentrations it allows estimation of the cellular phosphorylation potential and, by assuming that adenylate kinase is catalysing a near-equilibrium reaction, the cellular energy charge.

3.2 Studies of the control of oxidative phosphorylation *in vivo*

In skeletal muscle the determination of free ADP concentrations which were comparable with the K_m of oxidative phosphorylation for ADP in isolated mitochondria, strongly suggested that the level of ADP might control oxidative phosphorylation in this tissue. Chance and co-workers (22) distinguished between possible control by P_i and ADP by studying PFK deficient human muscle in which the level of P_i post-exercise was constant. This showed that phosphocreatine re-synthesis via oxidative phosphorylation in the post-exercise period must by controlled by the level of ADP. Radda and colleagues (7) measured the rate of phosphocreatine synthesis via oxidative phosphorylation in human muscles after exercise and showed that it was dependent on the ADP concentration in a hyperbolic manner. The K_m derived from these data for the ADP activation of oxidative phosphorylation was similar to that measured with isolated mitochondria *in vitro*. Measurements of the rate of phosphocreatine re-synthesis or ADP re-phosphorylation in human muscles post-exercise has been used to diagnose the presence of mitochondrial myopathy (7). In heart muscle, on the other hand, ADP does not seem to be the sole regulator of the rate of oxidative phosphorylation. ^{31}P NMR measurements on heart muscle *in situ* showed that over a wide-range of rate pressure products or work-loads the relative concentrations of phosphocreatine and ATP remained constant (23). Thus the calculated ADP concentration also remained constant and could not be responsible for the increased rate of myocardial oxidative phosphorylation and oxygen consumption that occurred in response to the increased work-load. In perfused hearts, however, mechanical output and oxygen consumption were found to be related to ADP levels by a Michaelis–Menten relationship when pyruvate was used as a substrate but not when glucose was the only carbon source in the perfusate. A comprehensive ^{31}P NMR study of the effect of different carbon sources on myocardial energetics showed that, in the presence of abundant oxygen, the rate of oxidative phosphorylation in this tissue was regulated by the availability of NADH and P_i as well as by ADP (24). Although not couched in the terms of Metabolic Control Analysis (25), this study underlines the general findings of that approach, i.e. there is no unique 'rate limiting' step and flux control in the tissue is distributed and dependent on conditions (Chapter 6).

3.3 Magnetization transfer measurements of ATP turnover

Magnetization transfer measurements *in vivo*, which were pioneered by the work of Brown *et al.* (26) on measurements of flux between P_i and ATP in *Escherichia coli*, provide a unique and powerful method for measuring rapid reaction rates in the cell. Although the number of fluxes that can be measured in the cell using this technique are relatively few, the flux between ATP and P_i is arguably one of the most important fluxes in the living cell. Measurements of ATP turnover from magnetization transfer measurements of P_i-> ATP flux (see *Figure 3*), coupled with simultaneous measurements of oxygen consumption can be used to determine the P/O ratio for oxidative phosphorylation *in vivo*, i.e. molecules of ATP synthesized per atom of oxygen consumed (Chapter 1).

In early measurements on yeast and perfused heart, measurements of P/O ratios which were greater than three were taken to indicate reversibility of mitochondrial ATP synthesis. However a number of subsequent studies showed that this ATP<->P_i exchange reaction in the cell was not due to the mitochondrial ATP synthase but to the coupled reaction catalysed by the glycolytic enzymes, glyceraldehyde-3-phosphate dehydrogenase (GAPDH) and phosphoglycerate kinase (PGK) (27–31). When this exchange reaction was eliminated, either by iodoacetate inhibition of GAPDH (28, 29, 31) or by lowering the cellular concentration of PGK using molecular genetic techniques (32), the measured P/O ratio in the NMR experiment was close to three. Thus there is no evidence for reversibility of the ATP synthase *in vivo*, as is required by the near-equilibrium hypothesis for the control of oxidative phosphorylation (33). There also appear to be no reports of lowered P/O ratios as a result of mitochondrial uncoupling other than that induced in the perfused rat heart by known uncouplers or high concentrations of fatty acids (34).

3.4 Substrate supply

The sensitivity of cellular energy status to the supply of nutrients, in particular oxygen, makes ^{31}P NMR a useful tool in monitoring tissue ischaemia and hypoxia (35). In heart, for example, ischaemia results in an initial loss of phosphocreatine and acidification of the muscle followed eventually by loss of ATP. These measurements can be made in the isolated perfused heart and non-invasively, in the intact animal, using spatial localization techniques (36). In humans, spatial localization techniques have been used to show a reduced phosphocreatine/ATP ratio in the failing hypertrophied heart, although this was thought to be due to a loss of creatine from the muscle rather than a change in the ADP/ATP ratio (37). There have been numerous ^{31}P NMR studies of tumours in which the aim has been to predict and monitor the response of the tumour to therapy. However studies on isolated tumour cells have sometimes failed to show a direct relationship between nutrient supply and energy status (38).

3.5 ^{31}P NMR measurements on cells

In reviewing the applications of ^{31}P NMR to the study of cellular energetics we have described NMR experiments which can be used in tissues ranging from perfused organs to the human brain. The non-invasive aspect of the technique is paramount in these investigations. How else, for example, could the pH in specific regions of the human brain be determined? Why then bother with measurements on isolated cells, given that they can be sampled readily and thus are much more amenable to more conventional biochemical analyses? There can be several reasons for wishing to use isolated cells. Tissues in an intact animal are inevitably heterogeneous and even with modern localization techniques the volume sampled may contain different cell types or the same cells experiencing different metabolic conditions. Measurements on isolated cells offer the prospect of studying an homogeneous cell preparation in which the supply of nutrients etc. can be precisely controlled. They may also aid in the interpretation of spectral changes observed in the same or similar cells in the intact animal in terms of the underlying biochemistry.

We have been using ^{31}P NMR in conjunction with molecular genetic manipulation of enzyme levels to study cellular bioenergetics in the yeast *Saccharomyces cerevisiae* (30, 39–42). We chose this organism as a model system to study as it is easy to grow in the large quantities needed for NMR experiments and because molecular genetic manipulation of the levels and properties of the enzymes involved in energy metabolism is relatively straight-forward (43). The protocols described below are based on this work. Where appropriate we also discuss how these protocols can be applied to other systems.

3.5.1 Cell immobilization

The inherent insensitivity of NMR means that in order to obtain spectra with good signal-to-noise ratios it is essential to maximize the viable cell density within the NMR probe. Although some studies have been performed on dense cell suspensions, in which the cells are stirred and oxygenated by bubbling oxygen through the suspension, these are essentially short-term experiments and most studies now employ some form of cell entrapment and perfusion. These systems allow maintenance of the cells in a metabolic steady state over the prolonged periods (hours) that are required for some NMR experiments. The different cell entrapment and perfusion systems available for microbial and mammalian cells have been reviewed recently (38).

In general we use the agarose gel thread entrapment system for cell immobilization, although for mammalian cells we have also used hollow-fibre cartridges (44). The agarose gel system, which was developed by Foxall and Cohen (45), is simple and easy to use and can be used with yeast, plant, and mammalian cells (38). The protocol described here (*Protocol 1*) is for 4 g wet weight of yeast, but can easily be scaled down and adapted for mammalian cells, which may only be available in much smaller quantities. Note that NMR experiments with less than 1 g wet weight of cells may prove difficult due to lack of signal.

Protocol 1. Cell immobilization for NMR experiments

Reagents

- Buffer A: 50 mM Mes buffer adjusted to pH 6.0 with NaOH, 2 mM KCl, 2 mM MgSO₄
- I litre of a stationary phase culture of S. *cerevisiae* grown on glucose[a]

A. *Yeast cell harvest*

1. Chill the culture on ice, and then harvest at 4°C by centrifugation at approximately 5000 *g* for 5–10 min.

2. Wash the cells with about 300 ml of chilled water and then with 100 ml of chilled buffer A. Re-sediment (10 min at 2000 *g*) in 50 ml centrifuge tube. Promptly decant the supernatant as the pellet will quickly reabsorb liquid making subsequent quantitation difficult.

B. *Cell immobilization*

This part of the protocol can be readily adapted for mammalian cells by substitution of buffer A with a buffer appropriate to the particular cell line being used.

1. Make up 10 ml of a 1.8% (w/v) solution of low gelling temperature agarose in buffer A, boil to dissolve, and then transfer to a water-bath at 37°C.

2. Weigh out 4 g of the yeast cell pellet from *Protocol 1A*, into a 50 ml disposable plastic centrifuge tube. Resuspend in chilled buffer and sediment as before. Transfer the tube containing 4 g of cells to a waterbath at 37°C. Allow 15 min for the temperature of the tube to equilibrate with the bath before proceeding.

3. Cut off the top half of the centrifuge tube and, using a 5 ml syringe, add 4 ml of the agarose solution to the yeast cell pellet. Gently mix by repeatedly drawing the mixture up into the syringe, while keeping the tube warm.

4. Push the end of a 30–40 cm piece of 0.5 mm i.d. polypropylene tubing (Portex PP-50, Portex Ltd.) on to a 0.6 mm i.d. 23 gauge syringe needle.

Protocol 1. *Continued*

5. Draw 1.5–2 ml of the cell/agarose mixture into a 2 ml syringe. Holding
   the syringe in the palm of the hand to keep it warm, carefully dry the
   Luer tip and then attach to the needle. Push the cell/agarose mixture
   to the end of the tubing and then chill it by briefly immersing the tube
   in an ice/water mix in a measuring cylinder. Extrude the solidified
   agarose gel thread into a 20 mm diameter NMR tube[b] filled with buffer
   A and chilled in an ice bucket. Stop the extrusion when you see non-
   solidified agarose appear at the tip of the tube. Chill the tubing again
   and repeat the process until the syringe is empty.

6. Repeat step 5 until all of the cell/agarose mixture has been used. Take
   care to dry the needle inlet and syringe tip between refills as a wet
   connection invariably comes loose.

7. Gently compact the cell/thread matrix into a volume of about 16 ml, e.g.
   to a depth of 65 mm in the 20 mm tube.[c] Two vortex plugs (Fluorochem
   Ltd.; Wilmad Glass Co.) are used to restrain the threads in the tube.
   These must be pre-drilled with holes to allow access of the inlet cannula
   and to allow the perfusate to flow out. A plug of cotton wool is placed
   over the first plug to prevent escape of the threads and is held in place
   by a second plug (see *Figure 5*).[d]

8. The yeast are typically perifused, at 25 ml/min, with 1 litre of buffer A
   containing 20 g glucose and sparged with pure O_2 (see *Figure 5*).

[a] This will provide typically more than 4 g of cells.
[b] For 1 g of cells or less use a 10 mm diameter NMR tube.
[c] Approx. 3 ml and a depth of 30 mm in a 10 mm diameter NMR tube.
[d] Note that most vortex plugs are made of Teflon and contract significantly when cooled.

3.5.2 Acquisition and quantitation of ^{31}P NMR spectra

The acquisition of spectra from live cells requires that the cells are maintained
in a viable state within the bore of the magnet (*Protocol 2*). The perfusion
system that we use in conjunction with the agarose gel immobilized cells is
illustrated in *Figure 5*. The pumps, water-baths, etc. will need to be maintained at a safe distance from the magnet (of the order of six to ten feet
depending on the magnet) and the perfusion circuit must be reliably leak-
proof.

Protocol 2. Acquisition and quantitation of ^{31}P NMR spectra of
cells

1. Using a standard sample, which has dimensions and an ionic strength
   comparable to that of the cell sample, tune the probe and determine
   the 90° pulse width.[a]

2. Place the cell sample in the probe and check the probe tuning. Shim on the water f.i.d. The ^1H NMR signal from water is sufficiently intense that this can be acquired using the ^{31}P-tuned coil. ^{31}P linewidths *in vivo* are inherently broad and a P_i linewidth of 20 Hz, at a field of 9.4 T, is acceptable.

3. Obtain fully relaxed spectra from yeast cells using a 90° pulse. This requires a pulse repetition rate of 5 sec or more, at fields of around 7–9 T. Use a data acquisition time of 0.5 sec or less and a spectral width of 10 kHz. Under these conditions usable spectra can be obtained at a field of 9.4 T (400 MHz ^1H frequency) in about 10 min (128 scans).

4. Quantitate spectra by comparison of the resonance intensities with that of a known standard. Estimate intracellular concentrations by multiplying the calculated concentration of the metabolite in the volume defined by the lower vortex plug by the ratio of this volume to the intracellular volume. The latter can be calculated by assuming that 1.67 g wet yeast contains 1 ml of cell water (46). This calculation assumes that the threads are homogeneously distributed over the volume defined by the vortex plug. Methylenediphosphonate (MDP) (Sigma) makes a useful standard in ^{31}P NMR as its resonance occurs a long way downfield from those of the common metabolites (see *Figure 4*). Prepare the standard as follows.

   (a) Use a buffered MDP solution (0.3 M MDP, 2 mM EDTA, 100 mM Hepes, pH 7.6) contained in a sealed 1.5 mm diameter capillary, 7 cm long, suspended from the vortex plug coaxially in the centre of the NMR tube. Calibrate the standard in the same diameter NMR tube used for the cells with a phosphate solution of known concentration and with an ionic strength similar to that of the cell preparation.

   (b) Acquire spectra at a 60 sec delay, which is fully relaxed for both the MDP and phosphate standard solution, and under the same acquisition conditions used for the cell experiments (5 sec repetition rate). The MDP resonance will be slightly saturated under the latter conditions but this saturated intensity can be directly related to the equivalent fully relaxed phosphate intensity, acquired at the 60 sec delay.

[a] For ^{31}P NMR work, 50 mM P_i, pH 7.0 is a suitable standard. Note that the ionic strength of the sample may have a significant effect on probing, tuning, and the pulse width. Carry out initial shimming of the probe also on this standard.

3.5.3 ^{31}P NMR measurements on cell extracts

The metabolites relevant to bioenergetic studies can be extracted from the cells with perchloric acid (*Protocol 3*) (*Warning*: organic perchlorates are

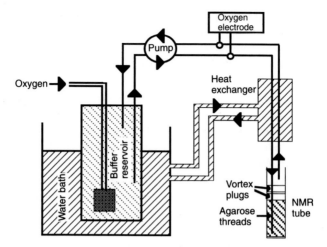

Figure 5. Schematic of a system used to perifuse immobilized cells in the bore of a spectrometer magnet. The heat exchanger sits in the bore of the magnet immediately above the probe and the NMR tube containing the cells.

highly explosive). Working with cell extracts has the advantage that spectral resolution can be markedly improved (see *Figure 4*). It is also possible to acquire signal over much longer periods of time and therefore to obtain better signal-to-noise ratios. To minimize hydrolysis, extracts are prepared from cells which have been rapidly frozen to liquid nitrogen temperatures. Note that inorganic phosphate can be artefactually released during this process and therefore can not be quantitated reliably. Using a high density perifused cell preparation (*Protocol 2*) has the advantage that it removes the requirement for the rapid centrifugation or filtration steps, needed with experiments on dilute cell suspensions.

Protocol 3. ^{31}P NMR measurements on cell extracts

A. *Preparation of extracts*

1. Remove the NMR tube from the cell perifusion apparatus as rapidly as possible.

2. Shake the threads out of the tube into a mortar pre-cooled with liquid nitrogen. Rapidly smear the threads and residual buffer around the mortar with a pre-cooled pestle.[a]

3. Grind the cells to a fine powder, adding more liquid nitrogen as required to keep the mortar and pestle cold. Then add, dropwise, 5 ml of cold 3 M perchloric acid. Grind the frozen perchloric acid in with the frozen cell powder.

---

4. Transfer the cell powder to a 50 ml disposable plastic centrifuge tube and allow the mixture to thaw before freeze–thawing a further two times in liquid nitrogen. Spin down the agarose/cell debris by centrifuging for 10 min at 2000 g. Keep the supernatant on ice. Resuspend the pellet with about 20 ml of water before re-centrifuging and pooling the supernatants.

5. Add 0.2 g Na_2EDTA to the supernatant. At this pH it will not dissolve. Neutralize the extract with cold 5 M K_2CO_3. Mix well to dissolve the EDTA and keep on ice. Spin down the potassium perchlorate precipitate and other insoluble matter by centrifuging at 2000 g for 10 min and transfer the supernatant to a clean tube.

6. Add about 2 g Chelex™ (Bio-Rad) resin to the extract and mix well. Leave to stand for about 5 min and then resuspend by gentle shaking. Spin down the Chelex resin (1 min at 2000 g) and decant the supernatant into a round-bottom flask. Wash the Chelex resin with about 20 ml water and add the washings to the flask.

7. Shell freeze the extract and freeze-dry. The dry extract can be stored at $-20°C$ prior to NMR measurements.

B. *Acquisition of NMR spectra*

The conditions used for acquisition will depend on whether quantitative information is required and which metabolites need to be measured. The pH of the sample is particularly important. We find that pH 8.0 gives good resolution of the signals of those metabolites which are of most interest.

1. Determine the ^{31}P 90° pulse width on a standard sample which has an high ionic strength similar to that of the resuspended extract.

2. Dissolve the lyophilized extract in 4 ml of a 50 mM triethanolamine buffer pH 8.0, containing 5 mM EDTA. Add 0.5 ml D_2O for a field frequency lock, adjust the pH to 8.0, and make up the volume to 5 ml with the triethanolamine buffer.

3. Remove insoluble material by centrifugation (10 min at 2000 g) and transfer 3–4 ml of the supernatant to a 10 mm diameter NMR tube. Include a chemical shift and concentration reference standard. Use a 40 mM solution of MDP contained in a coaxial capillary suspended from a vortex plug or, if the sample does not contain phosphocreatine (as is usually the case in yeast) the sample can be spiked with a known concentration of phosphocreatine. Place the sample in the magnet, tune the probe, and shim.

4. Acquire spectra with ^1H decoupling, using the lowest power compatible with complete decoupling. With these high ionic strength samples heating due to the decoupler can be a problem. Keep the sample temperature constant.[b]

Protocol 3. *Continued*

5. Resonance assignments can often be made by reference to the literature. However this can be misleading as many of the ^{31}P resonances
shift with pH and other factors.[c]

6. Quantitate by comparison with the resonance intensity of a standard
of known concentration (see *Protocol 2*).[d]

[a] The cells freeze rapidly on contact with the porcelain therefore there is no need to have
liquid nitrogen in the mortar when the cells are added.

[b] We normally use a temperature of 30°C. For metabolite quantitation it is best to acquire
fully relaxed spectra. However the spectra can be acquired more rapidly and 'saturation factors'
determined by comparing these spectra with a fully relaxed spectrum. At a field of 7 T fully
relaxed spectra can be acquired with 60° pulses and a 4 sec repetition rate. Under these
conditions spectra with good signal-to-noise can be acquired in a few hours.

[c] Definitive assignments can be made using a number of different strategies for example:
  (a) Add enzyme(s) which degrade specific metabolites in the extract.
  (b) Adding a small quantity of the candidate compound, typically enough to double the
     intensity of the existing signal. It is sometimes necessary to examine the titration
     behaviour of the resonance in question before and after adding the candidate compound.

[d] On most modern spectrometers there is peak integration software to do this. Take care to
ensure that the correct spectral baseline is used. Overlapping peaks can be deconvoluted by
fitting Lorentzian lines to the observed peaks. Verify quantitation of partially relaxed spectra
by adding a known concentration of the genuine compound and measuring the increase in
peak intensity.

3.5.4 Saturation transfer measurements on cells

The basic principles of the method are described in the text (sections 2.2 and
3.3). The NMR acquisition conditions described in *Protocol 4* are appropriate
for the immobilized yeast cell system described above (*Protocol 2*), although
they may be applied to other systems with minor modifications.

Protocol 4. Saturation transfer measurements of P_i-> ATP flux

1. Acquire an initial spectrum to determine the frequency offsets of the
P_i and ATP γ-phosphate resonances.

2. Saturate the ATP resonance by applying a low power pulse at its
resonance frequency in the 5 sec delay between the end of the acquisition period and the next 90° pulse.[a]

3. After 8 or 16 scans, switch the saturating field to a frequency downfield of
the P_i resonance equal to the frequency difference between the resonances of P_i and the γ-phosphate resonance of ATP. This is the control
irradiation. Collect another 8 or 16 scans and repeat step 1 until a total of
between 256–512 scans have been acquired for each spectrum. This
interleaving of the irradiations and the storage of the resultant spectra
should be handled automatically by the spectrometer software. Comparison of the P_i peak intensities in these two spectra gives the ratio M_z/M_0.

4. Perform an inversion recovery T_1 measurement on the P_i resonance in the presence of selective saturation of the γ-phosphate resonance of ATP.

   (a) Measure the P_i z magnetization at 10–15 different delay times (τ) following its inversion, using the pulse sequence 180°–τ–90°– acquire. The delay time should be varied between approximately 50 msec and 5 sec. The spectra at each delay can be collected in blocks of 16 scans. Four cycles through the delay list will give a total of 64 scans for each spectrum.

   (b) Measure the P_i peak intensities and fit them to the function:

   $$M_z = M_0 \, (1 - 2e^{-at})$$

   where a is equal to $1/T_{1p} + k$. T_{1p} is the intrinsic T_1 of the P_i resonance and k is the first order rate constant for the exchange between P_i and ATP.

5. Combine the value for a with the measurement for M_z/M_0 from step 3 using Equation 11 to determine k. Multiplication of this rate constant (k) by the intracellular P_i concentration gives the flux.

[a] The power of this saturating pulse should be the minimum that is required to achieve saturation of the ATP signal. (In practice this corresponds to a γB_2 field of between 100–200 rad/sec.) Excessive power will lead to non-specific saturation. This saturating field can be generated in a number of different ways, depending on the type and age of the spectrometer (consult the spectroscopist responsible for the instrument), for example by using a second frequency source and pulse amplifier, or by using the main frequency source and switching the power of the main pulse amplifier, or by using hard pulses on this channel in the DANTE sequence.

3.5.5 Investigation of cellular energetics using ^{31}P NMR and molecular genetics

There is an argument that we have reached the limits of reductionism in terms of what it can tell us about the control of metabolism *in vivo* (47), i.e. the breaking open of the cell and deducing the mechanisms of metabolic control from a study of its component parts. There is a need, therefore, for developing new methods for studying control *in vivo*. The combination of genetic manipulation of cellular enzyme levels followed by characterization of the consequent changes in metabolic fluxes and metabolite levels, using non-invasive NMR techniques and more conventional biochemical methods, provides a powerful approach for studying the control of metabolism *in vivo*. We have been using this approach in the yeast, *S. cerevisiae*, to study control of glycolytic flux. This has included a study of the effects of 6-phosphofructo-1-kinase overexpression on ATP generation via glycolysis and mitochondrial oxidative phosphorylation (41), and a study of the effect of phosphoglycerate kinase under- and overexpression on glycolytic flux and ATP<->P_i exchange, measured using magnetization transfer techniques (30).

^{31}P NMR spectra of most biological systems give no direct information on the phosphorylation potential as ADP is below the threshold of detection. However in those systems possessing creatine kinase, such as muscle and brain, the free cytosolic ADP concentration can be estimated from the phosphocreatine/ATP ratio. This measurement can be made in yeast by introducing creatine kinase into the cell using molecular genetic techniques and incubating the cells with the enzyme's substrate, creatine (39). Although this experiment did not work well using creatine, more recent work has indicated that it will work with the creatine analogue, cyclocreatine.

Thus by using molecular genetic techniques in combination with non-invasive NMR measurements we can not only derive information on the role played by a specific enzyme on flux control *in vivo*, we can also engineer the cell so that the ^{31}P NMR experiment will give more information. These types of experiments are not restricted to yeast. For example creatine kinase has been expressed in the livers of transgenic mice so that ^{31}P NMR can be used to estimate the free ADP concentration in this tissue (21) and the enzyme has been deleted in the muscles of mice, using gene targeting experiments, in ^{31}P NMR studies of muscle function (48).

In summary, the perfused yeast cell/NMR system described here is a relatively simple, easily controlled, and readily manipulable system which promises to be a useful tool in understanding the control of energy metabolism *in vivo*.

Acknowledgements

We would like to thank the Wellcome Trust, the SERC, and the Royal Society for supporting the research in KMBs laboratory.

References

1. Cerdan, S. and Seelig, J. (1990). *Annu. Rev. Biophys. Biophys. Chem.*, **19**, 43.
2. Radda, G. K. (1992). *FASEB J.*, **6**, 3032.
3. Smith, G. A., Hesketh, R. T., Metcalfe, J. C., Feeney, J., and Morris, P. G. (1983). *Proc. Natl. Acad. Sci. USA*, **80**, 7178.
4. Moon, R. B. and Richards, J. H. (1973). *J. Biol. Chem.*, **248**, 7276.
5. Hoult, D. I., Busby, S. J. W., Gadian, D. G., Radda, G. K., Richards, R. E., and Seeley, P. J. (1974). *Nature*, **252**, 285.
6. Ackerman, J. J. H., Grove, T. H., Wong, G. G., Gadian, D. G., and Radda, G. K. (1980). *Nature*, **283**, 167.
7. Radda, G. K. (1986). *Science*, **233**, 640.
8. Gadian, D. G. (1982). *Nuclear magnetic resonance and its applications to living systems*. Clarendon Press, Oxford.
9. Derome, A. E. (1987). *Modern NMR techniques for chemistry research*. Pergamon Press, New York.

10. Sanders, J. K. M. and Hunter, B. K. (1987). *Modern NMR spectroscopy.* Oxford University Press, New York.
11. Rao, B. D. N., Cohn, M., and Scopes, R. K. (1978). *J. Biol. Chem.*, **253**, 8056.
12. Brindle, K. M. (1988). *Prog. Nucl. Magn. Reson. Spectrosc.*, **20**, 257.
13. Brindle, K. M., Blackledge, M. J., Challiss, R. A. J., and Radda, G. K. (1989). *Biochemistry*, **28**, 4887.
14. Arnold, D. L., Matthews, P. M., and Radda, G. K. (1984). *Magn. Reson. Med.*, **1**, 307.
15. Garlick, P. B., Soboll, S., and Bullock, G. R. (1992). *Nucl. Magn. Reson. Biomed.*, **5**, 29.
16. Murphy, E., Gabel, S. A., Funk, A., and London, R. E. (1988). *Biochemistry*, **31**, 526.
17. Masson, S. and Quistorff, B. (1992). *Biochemistry*, **31**, 7488.
18. Hutson, S. M., Berkich, D., Williams, G. D., LaNoue, K. F., and Briggs, R. W. (1989). *Biochemistry*, **281**, 4325.
19. Wallimann, T., Wyss, M., Brdiczka, D., Nicolay, K., and Eppenberger, H. M. (1992). *Biochem. J.*, **281**, 21.
20. Lawson, J. W. R. and Veech, R. L. (1979). *J. Biol. Chem.*, **254**, 6528.
21. Koretsky, A. P., Brosnan, M. J., Chen, L., Chen, J., and Van Dyke, T. (1990). *Proc. Natl. Acad. Sci. USA*, **87**, 3112.
22. Chance, B., Eleff, S., Bank, W., Leigh, J. S., and Warnell, R. (1982). *Proc. Natl. Acad. Sci. USA*, **79**, 7714.
23. Balaban, R. S., Kantor, H. L., Katz, L. A., and Briggs, R. W. (1986). *Science*, **232**, 1121.
24. From, A. H. L., Zimmer, S. D., Michurski, S. P., Mohanakrishnan, P., Ulstad, V. K., Thoma, W. J., and Ugurbil, K. (1990). *Biochemistry*, **29**, 3731.
25. Fell, D. A. (1992). *Biochem. J.*, **286**, 313.
26. Brown, T. R., Ugurbil, K., and Shulman, R. G. (1977). *Proc. Natl. Acad. Sci. USA*, **74**, 5551.
27. Brindle, K. M. and Radda, G. K. (1987). *Biochim. Biophys. Acta*, **928**, 45.
28. Campbell-Burk, S. L., Jones, K. A., and Shulman, R. G. (1987). *Biochemistry*, **26**, 7483.
29. Kingsley-Hickman, P. B., Sako, E. Y., Mohanakrishnan, P., Robitaille, P. M. L., From, A. H. L., Foker, J. E., and Ugurbil, K. (1987). *Biochemistry*, **26**, 7501.
30. Brindle, K. M. (1988). *Biochemistry*, **27**, 6187.
31. Mitsumori, F., Rees, D., Brindle K. M., Radda, G. K., and Campbell, I. D. (1988). *Biochim. Biophys. Acta*, **969**, 185.
32. Brindle, K. M., Fulton, A. M., and Williams, S.-P. (1994). In *Nuclear magnetic resonance in physiology and medicine* (ed. R. J. Gillies), pp. 237–62. Academic Press Inc., San Diego.
33. Erecinska, M. and Wilson, D. F. (1982). *Membr. Biol.*, **70**, 1.
34. Kingsley-Hickman, P. B., Sako, E. Y., Ugurbil, K., From, A. H. L., and Foker, J. E. (1990). *J. Biol. Chem.*, **265**, 1545.
35. Gadian, D. G. and Radda, G. K. (1981). *Annu. Rev. Biochem.*, **50**, 69.
36. Bottomely, P. A., Herfkens, R. J., Smith, L. S., Brazzamano, S., Blinder, R., Hedlund, L. W., Swain, J. L., and Redington, R. W. (1985). *Proc. Natl. Acad. Sci. USA*, **82**, 8747.

37. Conway, M. A., Allis, J., Ouwerkerk, R., Niioka, T., Rajagopalan, B., and Radda, G. K. (1991). *Lancet*, **338**, 973.
38. Szwergold, B. S. (1992). *Annu. Rev. Physiol.*, **54**, 775.
39. Brindle, K., Braddock, P., and Fulton, S. (1990). *Biochemistry*, **29**, 3295.
40. Brindle, K. M., Davies, S. E. C., and Williams, S.-P. (1991). *Biochem. Soc. Trans.*, **19**, 997.
41. Davies, S. E. C. and Brindle, K. M. (1992). *Biochemistry*, **31**, 4729.
42. Williams, S.-P., Fulton, A. M., and Brindle, K. M. (1993). *Biochemistry*, **32**, 4895.
43. Guthrie, C. and Fink, G. R. (ed.) (1991). *Methods in enzymology*, Vol. 194. Academic Press, San Diego.
44. Callies, R., Jackson, M. E., and Brindle, K. M. (1993). *Bio/Technology*, **12**, 75.
45. Foxall, D. L. and Cohen, J. S. (1983). *J. Magn. Reson.*, **52**, 346.
46. Gancedo, J. M. and Gancedo, C. (1973). *Biochemie*, **55**, 205.
47. Fraenkel, D. G. (1992). *Annu. Rev. Genet.*, **26**, 150.
48. van Deursen, J., Heerschap, A., Oerlemans, F., Ruitenbeek, W., Jap, P., ter Laak, H., and Wieringa, B. (1993). *Cell*, **74**, 621.

9

Isolation and characterization of photosynthetic reaction centres from eukaryotic organisms

MICHAEL C. W. EVANS, BEVERLY J. HALLAHAN,
JONATHAN A. HANLEY, PETER HEATHCOTE,
NICOLA J. GUMPEL, and SAUL PURTON

1. Introduction

Photosynthetic reaction centres are the site of solar energy conversion in biological systems. The basic organization of photosynthetic systems is the same in all chlorophyll-containing organisms. Light is absorbed by a large array of chlorophyll molecules and the energy transferred by random processes through the array to a reaction centre chlorophyll, which undergoes photochemical oxidation transferring an electron to an acceptor molecule. The chlorophyll array is highly organized with the pigments bound to a range of proteins which modulate the chlorophyll energy levels and provide an organization which facilitates energy transfer to the reaction centre. The light harvesting protein complexes surround the reaction centre forming a photosynthetic unit. The photosynthetic units are in turn bound in a specific membrane system, the thylakoid. In eukaryotic organisms the thylakoids are in turn limited to a specialized organelle, the chloroplast. The high chlorophyll to reaction centre concentration makes optical analysis of the reaction centre extremely difficult and also limits the concentration of reaction centres which can be obtained for spectroscopic analysis. The presence of two types of reaction centre in oxygen evolving organisms further complicates analysis. The successful development of models of reaction centre structure and mechanism has depended on the development of techniques to isolate and characterize the reaction centres.

Procedures for the separation of the two photosystems were developed 25 years ago and these still form the basis of most preparations. Photosystem II (PSII) has been extensively purified with the isolation of a number of 'core complexes' which retain the bound electron transport system and oxygen evolution capability, and reaction centre preparations (containing the polypeptides D1, D2, and the cytochrome b_{559} heterodimer) which are analogous to purple bacterial reaction centres. However, the reaction centre preparations are very unstable, lack the electron accepting quinone complex, and have an altered electron donor environment.

Photosystem I (PSI) has also been purified to some extent. Its structure is different from that of PSII and the purple bacterial complex, with a large number of light harvesting chlorophyll molecules associated with the reaction centre polypeptides, and membrane bound electron transport components bound by low molecular weight polypeptides. The most frequently used detergent, Triton X-100, affects energy transfer and possibly electron transfer pathways, while detergents such as digitonin separate but do not otherwise purify the reaction centres.

It seems likely that the prosthetic groups involved in electron transfer in both photosystems have been fully identified. However the failure to obtain pure native reaction centres from eukaryotic organisms means that our knowledge of polypeptide composition, three-dimensional structure, and kinetic behaviour of the reaction centres is much less precise than in purple bacteria. Equally the application of molecular genetic techniques is at a much earlier stage and the properties of mutants of algae, such as *Chlamydomonas reinhardtii*, lacking either reaction centre or the light harvesting complexes have not yet been fully investigated.

In this chapter we describe procedures for the isolation and purification of PSI and PSII from eukaryotic organisms. These procedures have been developed essentially for work on spinach. Our experience is that they work equally well with only minor modification for other higher plants and algae having similar chloroplast structures. More extensive modification may be required for algae in which the chloroplasts do not have grana stacks. The techniques have been used to provide relatively large amounts of material for biophysical analysis, we have not attempted to include the huge range of procedures which have been used to provide small amounts of material for polypeptide analysis by gel electrophoresis. We also describe the characterization of these preparations by a range of techniques, some of which are technically specialized and involve expensive equipment. Our main objective is to describe the sample preparation procedures, for at least some of the techniques it is only sensible to suggest that measurements should be carried out in collaboration with specialist laboratories. These techniques can often be applied to intact organisms and their use to screen and analyse photosynthetic mutants of *C. reinhardtii* are described.

The purification of these preparations can be followed by a number of criteria, commonly reaction centre to chlorophyll ratios, chlorophyll *a/b* ratios, rates of oxygen evolution, or NADP reduction may be used. No single measure can be used, improvements in reaction centre to chlorophyll ratios are usually accompanied by loss of overall activity. Increased rates of oxygen evolution may actually reflect damage to the preparation increasing accessability to external electron acceptors rather than increased purity of the preparation. The most appropriate criterion must therefore be selected for the immediate experimental purpose.

2. Purification of PSII

The isolation of intact chloroplasts, thylakoid membranes, and reaction centre core complexes has been covered in an earlier edition (1). This section is intended to complement and update the previously described methods.

2.1 Thylakoid membranes

The first step in the preparation of thylakoid membranes is the isolation of chloroplasts. The thylakoids are then released by osmotic lysis of the chloroplast envelope. All stages should be carried out at 0–4 °C unless otherwise stated. In order to protect the PSII particles from photodamage and obtain maximum rates of oxygen evolution it is recommended that steps following the purification of chloroplasts are performed under subdued lighting.

Protocol 1. Preparation of thylakoid membranes

Equipment and reagents

- A crate of market bought spinach or several trays of 14-day-old pea seedlings
- A laboratory blender
- Grinding medium: 0.33 M sorbitol, 0.2 mM $MgCl_2$, 20 mM Mes–NaOH pH 6.5, pre-cooled to ice temperature–add sodium

- ascorbate to a final concentration of 5 mM just prior to use
- 5 mM $MgCl_2$
- Resuspending medium: 25 mM NaCl, 5 mM $MgCl_2$, 20 mM Mes–NaOH pH 6.3

Method

1. Wash the spinach leaves. Remove and discard the large leaf stems and ribs. Harvest pea seedlings with scissors just before use.

2. Homogenize the leaves in grinding medium for 15 sec in a Waring blender.

3. Filter the slurry through eight layers of muslin and centrifuge the filtrate at 3000 *g* for 5 min to pellet the chloroplasts.

4. Resuspend the pellet in the hypotonic 5 mM $MgCl_2$ solution for 60 sec to osmotically shock the chloroplasts. Add an equal volume of double strength grinding medium and centrifuge at 3000 *g* for 20 min.

5. Resuspend the thylakoid pellet in resuspending medium. Determine the chlorophyll concentration as described in *Protocol 9*. The chlorophyll *a* to chlorophyll *b* ratio should be approximately three.

2.2 PSII particles

PSI and PSII complexes are solubilized from the thylakoid membranes using the detergent Triton X-100 and the complexes separated by centrifugation. The method described (*Protocol 2*) is that of Ford and Evans (2).

Protocol 2. Isolation of PSII particles

Reagents

- Resuspending medium containing 0.33 M sucrose or 20% glycerol
- 20% (v/v) Triton X-100

Method

1. Incubate the thylakoid membrane preparation on ice in darkness for 2 h to allow lateral migration and segregation of PSI and PSII complexes.

2. Solubilize the thylakoids by adding resuspending medium and Triton X-100 so that the final chlorophyll concentration is 2 mg/ml and the final detergent concentration is 5% (w/v). Mix gently by inversion and leave on ice for 25 min.

3. Centrifuge at 40 000 g for 30 min. The PSII particles are now located in the pellet and the PSI complexes in the supernatant.

4. Decant the supernatant and resuspend the PSII particles in resuspending medium containing the cryoprotectant.

5. Measure the chlorophyll concentration and determine the chlorophyll *a* to *b* ratio (see *Protocol 9*). This ratio should have fallen to about two. The PSII particles may now be stored at 77 K.

Assays for oxygen evolution may be performed in the presence of the electron acceptor 2,6 dimethylbenzoquinone (DMBQ) with an oxygen electrode, the use of which has been extensively covered in an earlier edition (3). Oxygen evolution for a typical preparation will be in the range 300–600 μmol O_2/mg chl/h.

2.3 PSII core preparations

The PSII particles obtained by the above method may be further purified by detergents, resulting in core particles (*Protocol 3*). Treatment of PSII particles with octylglucopyranoside (OGP) removes the light harvesting complex (LHCII), leaving a core particle that retains D1, D2, CP47, CP43, CP29 the cytochrome b_{559} polypeptides, the 22 and 10 kDa subunits, and the 33 kDa extrinsic polypeptide. The core particles also retain the manganese ions associated with water oxidation. The use of *n*-heptylthioglucoside (HTG) produces a similar loss of light harvesting polypeptides but all three extrinsic polypeptides (17, 23, and 33 kDa) are retained (*Protocol 4*).

Protocol 3. Preparation of OGP (octylglucopyranoside) core particles

Reagents

- Buffer P3-1: 0.4 M sucrose, 10 mM NaCl, 50 mM NaHCO$_3$, 50 mM Hepes–NaOH pH 7.5
- Buffer P3-2: 1 M sucrose, 0.8 M NaCl, 50 mM NaHCO$_3$, 75 mM octylglucopyranoside, 50 mM Hepes–NaOH pH 7.5
- Buffer P3-3: 1 M sucrose, 0.4 M NaCl, 50 mM NaHCO$_3$, 50 mM Hepes–NaOH pH 7.5
- Buffer P3-4: 50 mM NaCl, 50 mM NaHCO$_3$, 50 mM Hepes–NaOH pH 7.5
- Buffer P3-5: 15 mM NaCl, 5 mM MgCl$_2$, 50 mM NaHCO$_3$, 20 mM Hepes–NaOH pH 7.5, 20% (v/v) glycerol

Method

1. Resuspend PSII particles prepared as above in P3-1 to give a chlorophyll concentration of 2.5 mg/ml.

2. Add an equal volume of P3-2. Mix gently and incubate on ice for 12 min.

3. Dilute the digestion mixture with 2 vol. of P3-3. Centrifuge at 40 000 *g* for 30 min.

4. Dialyse the supernatant against P3-4 for 60 min, then centrifuge at 40 000 *g* for 60 min.

5. Resuspend the pellet in P3-5. The determined chlorophyll *a* to chlorophyll *b* ratio of the OGP core particles should show an increase to around ten as the chlorophyll *b* content is reduced with the removal of LHCII. The OGP core particles may now be stored at 77 K.

Protocol 4. Preparation of HTG (*n*-heptylthioglucoside) core particles

Reagents

- Buffer P4-1: 1 M sucrose, 40 mM MgCl$_2$, 10 mM NaCl, 20 mM NaHCO$_3$, 2% (w/v) *n*-heptylthioglucoside, 40 mM Hepes–NaOH pH 7.5
- Buffer P4-2: 0.5 M sucrose, 20 mM NaHCO$_3$, 40 mM Hepes–NaOH pH 7.5
- Buffer P4-3: 20 mM NaHCO$_3$, 40 mM Hepes–NaOH pH 7.5
- Buffer P4-4: 10 mM NaCl, 20 mM NaHCO$_3$, 1 mM EDTA, 40 mM Hepes–NaOH pH 7.5
- Buffer P4-5: 10 mM NaCl, 5 mM MgCl$_2$, 20 mM NaHCO$_3$, 40 mM Hepes–NaOH pH 7.5, 20% (v/v) glycerol

Method

1. Resuspend PSII particles obtained from *Protocol 2* to 2 mg/ml chlorophyll in P4-1.

Protocol 4. *Continued*

2. Incubate on ice for 12 min then add 1.5 vol. of P4-2.
3. Centrifuge at 40 000 *g* for 45 min. Recover the supernatant and mix with 2 vol. of P4-3.
4. Centrifuge at 40 000 *g* for 60 min. Resuspend the pellet in P4-4.
5. Pellet the HTG core particles by centrifugation at 40 000 *g* for 10 min. Finally resuspend in P4-5.

HTG core particles (and OGP core particles) may be prepared in the absence of bicarbonate ions, at pH 6. Mes is used instead of Hepes as the buffering agent and bicarbonate omitted from the buffers. This results in a HTG core preparation with an increased rate of oxygen evolution when using DMBQ as the electron acceptor. However, bound bicarbonate is lost from the acceptor side of PSII as evidenced by the presence of the modified Q_{A-} Fe^{2+} 1.8 EPR signal (see section 3.2).

2.4 Reaction centre core particles

Reaction centre core particles may be obtained by treatment of OGP core particles with dodecylmaltoside (4) followed by separation on S-sepharose and Q-sepharose columns. This results in a complex consisting of D1, D2, cytochrome b_{559}, and CP47. A second method (5) described below (*Protocol 5*) yields a D1/D2/cyt b_{559} complex. The method involves the digestion of PSII particles with Triton X-100 followed by ion exchange chromatography. It should be noted that the reaction centre core particles prepared by both methods also contain small PSII polypeptides of less than 5 kDa.

The first step in the preparation of the D1/D2/cyt b_{559} complex is the removal of the 17, 23, and 33 kDa extrinsic polypeptides by washing with Tris buffer. This procedure is carried out under room light since the slow turnover of the 'S' states of the oxygen evolving complex increases the chances of release of the extrinsic polypeptides. The inclusion of *ortho*-phenanthroline throughout the purification results in an increased yield of reaction centre core particles. This may be because the *ortho*-phenanthroline facilitates the release of quinones and/or the non-haem iron. Trition is substituted by dodecylmaltoside during elution of the reaction centre core particles to increase the stability of the complexes.

Protocol 5. Isolation of the b_{559} complex

Equipment and reagents

- 30 cm^3 DEAE-Fractogel column pre-equilibrated with 30 mM NaCl, 1 mM *ortho*-phenanthroline, 50 mM Tris–HCl pH 7.2, 0.2% Triton X-100
- Ultrafiltration concentrator system with a 100 kDa cutoff membrane
- Elution buffer: 1 mM *ortho*-phenanthroline, 4 mM dodecylmaltoside, 50 mM Tris–HCl pH 7.2

Method

1. Dilute PSII particles from *Protocol 2* tenfold with 1 M Tris–HCl pH 8.8. Incubate on ice for 60 min under room light.

2. Centrifuge at 40 000 *g* for 30 min at 4°C. Resuspend the resulting pellet in 50 mM Tris–HCl pH 7.2. Add 1 mM *ortho*-phenanthroline and Triton X-100 such that the final detergent concentration is 4% (w/v) and the final chlorophyll concentration is 1 mg/ml. Stir gently on ice, in darkness for 90 min.

3. Centrifuge at 40 000 *g* for 30 min. Load the supernatant on to the pre-equilibrated column and wash with approximately 20 column volumes of the equilibration buffer. The eluate should now be colourless.

4. Elute the reaction centres from the column using a salt gradient in elution buffer. Increase the gradient in 30 mM steps up to 120 mM.

5. Concentrate the reaction centre core particles using the ultrafiltration concentrator. For storage at 77 K add glycerol to 20% (v/v).

Elution of the reaction centre core particles from the column may be followed by measuring the absorption spectrum of the eluted fractions. A peak at around 417 nm is due to phaeophytin *a* and the Soret band of oxidized cytochrome b_{559}. The peak at around 432 nm is the Soret band of chlorophyll *a*. The enrichment of the D1/D2/cyt b_{559} complex may be followed by observing an increasing ratio between these two bands (417 nm/432 nm). The chlorophyll *a* absorption peak at 676 nm may also be used as an indicator.

3. EPR analysis of photosystem II

EPR samples are prepared in silica tubes (3 mm inner diameter, 4 mm outer diameter). Tubes should have matched diameters if any quantitative results are required. Molecular oxygen is paramagnetic and the presence of O_2 in frozen samples can cause major distortion of the EPR spectra. Samples should always be prepared in the absence of oxygen. This is most easily achieved by preparing all samples under a stream of nitrogen blown over the surface of the sample in the EPR tube, using a long syringe needle or equivalent metal tube. However, this is not essential for most PSII EPR samples. A sample length of 1–2 cm is required for most EPR spectrometers, requiring a sample volume of 0.2–0.3 ml in standard tubes. The sample should be introduced into the tube with a syringe, preferably using a plastic cannula as a needle. Additions of oxidant, reductant, etc. can be made with μl syringes with long needles, using the needle to mix the sample after addition.

3.1 Signal II

Two redox active tyrosine residues are associated with the donor side of photosystem II. Electron transfer between the oxygen evolving complex and

P680 is mediated by tyrosine Y_Z. The second tyrosine, Y_D does not appear to directly participate in oxygen evolution. The paramagnetic species $Y_Z\cdot$ and $Y_D\cdot$ give rise to a characteristic EPR spectrum termed signal II, the different forms of signal II being distinguished by their decay kinetics at room temperature. Signal II very fast (vf) which decays on a microsecond time-scale at physiological temperatures arises from $Y_Z\cdot$. When the manganese cluster is destroyed or uncoupled from Y_Z, $Y_Z\cdot$ gives rise to a similar signal, signal II fast (f) with millisecond reduction kinetics.

Signal II slow (s) has much slower decay kinetics and arises from $Y_D\cdot$. It may be observed in chloroplast membranes or photosystem II preparations. Up to 75% of PSII centres in a sample dark adapted at 273 K will demonstrate a dark stable form of signal II, arising from oxidized $Y_D\cdot$. To observe $Y_D\cdot$ in the PSII centres you will need:

- commercial EPR X-band spectrometer fitted with a liquid helium cryostat!
- PSII particles as prepared in section 2.2 resuspended to a concentration of 5 mg chl/ml
- calibrated quartz EPR tubes
- 650–1000 W light source
- silvered Dewar containing liquid nitrogen

Transfer 0.3–0.4 ml of the sample to an EPR tube. Position the EPR tube over the top of the Dewar so that it is cooled by the evaporating nitrogen and illuminate the sample for 1 min by holding the light source approximately 1 m away from the tube. Rotate the tube continuously to prevent sample heating. Gently lower the tube into the liquid nitrogen, continuing the illumination until the sample is frozen.

The EPR signal of $Y_D\cdot$ can be observed over a wide-range of temperatures, from below 10 K to room temperature. To obtain the best signal-to-noise ratio, a temperature of 15 K with a microwave power setting of 1 μW is recommended. Other EPR conditions are given in *Figure 1*. The signal has a g value of 2.0046, a linewidth of 1.9 mT, and partially resolved hyperfine peaks at about 0.5 mT apart. The area of the signal from $Y_D\cdot$ obtained by double integration may be assumed to represent one spin per PSII reaction centre for quantitation of other free radical signals.

3.2 The iron quinone signals

Two plastoquinones, Q_A and Q_B are associated with the acceptor side of PSII. The primary quinone Q_A accepts only one electron under normal conditions, forming the semi-quinone radical Q_A-. Q_B receives two electrons from Q_A before dissociating from PSII in the quinol form. Stable reduction of Q_A can be achieved by either chemical reduction, freezing under illumination (Q_B becomes doubly reduced), illumination in the presence of DCMU which inhibits electron transport beyond Q_A, or by illumination at temperatures below 230 K where the transfer of electrons to Q_B is thermodynamically

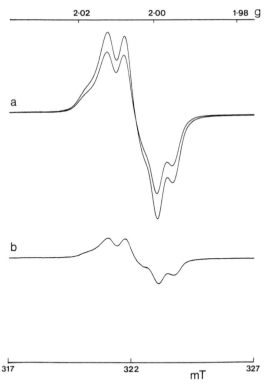

Figure 1. EPR spectrum of Y_D· in PSII particles. Difference spectrum (b) shows the signal induced by 77 K illumination (a, outer line) of a sample stored in the dark at 77 K for one week (a, inner line) following freezing under illumination. EPR conditions: temperature, 15 K; microwave power, 1 μW; modulation width, 0.2 mT.

unfavourable. This region of PSII is also influenced by the non-haem iron which is bound by ligands from both of the reaction centre polypeptides, D1 and D2, and the bicarbonate ion which probably provides at least one ligand to the iron. A semi-quinone would be expected to have a narrow Gaussian $g = 2$ EPR signal typical of an organic radical, but in the PSII reaction centre the non-haem iron exerts a magnetic influence over Q_A and Q_B so that the main resonances are observed at g values less than two. Variations in the g value and lineshape of the signal are dependent on the presence or absence of bicarbonate, other anions, or herbicides. With bicarbonate present the native EPR signal at $g = 1.9$ is obtained. This is the Q_A- Fe^{2+} signal normally observed in PSII particles as prepared in section 2.2.

To see the $g = 1.9$ signal in PSII particles, transfer a sample to an EPR tube. Dark adapt the sample on ice for 60 min then freeze in liquid nitrogen in the dark. Prior to examination by EPR, illuminate the sample at 77 K using a lamp (650 W) positioned over a silvered Dewar or from the side using an

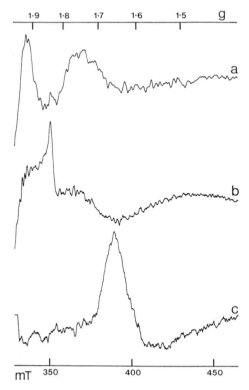

Figure 2. The iron quinone signals of photosystem II. (a) $g = 1.9$ signal from spinach PSII particles, (b) $g = 1.8$ signal from spinach HTG core particles, (c) $g = 1.6$ signal from *Phormidium laminosum* PSII particles. All samples were illuminated at 77 K. EPR conditions: temperature, 4.5 K; microwave power, 10 mW; modulation width, 1.25 mT.

unsilvered Dewar. Illuminate the sample for 10 min. The signal is best observed at low temperatures (4–15 K) and high powers (10–20 mW). Other EPR conditions are given in *Figure 2*.

When bicarbonate is displaced from the reaction centre by conditions of low pH or by treatment with 50 mM formate, the $g = 1.8$ EPR signal is seen. The two types of $Q_{A^-} Fe^{2+}$ signal ($g = 1.8$, $g = 1.9$) can be clearly observed in OGP core particles prepared at pH 6.0 and pH 7.5 (+ bicarbonate), respectively (see section 2.3). Sample preparation is as above except that a chlorophyll concentration of 2 mg/ml is sufficient.

A third EPR signal associated with the iron quinone may be observed in PSII particles and can be used as a marker for the presence of Q_B semi-quinone. This signal has a peak at $g = 1.63$ and arises from the $Q_{A^-} Fe^{2+} Q_{B^-}$ interaction. The $g = 1.6$ signal is seen only in the presence of bound bicarbonate. To observe this signal illuminate PSII particles at 4 °C for 1 min, then dark adapt for 45 min on ice and freeze to 77 K. Illuminate the sample at

77 K for 5 min, as described above. Measure the signal at 4–5 K with a microwave power of 10 mW and modulation width 1.25 mT.

The non-haem iron itself displays an EPR signal if oxidized to the ferric state, either by potassium ferricyanide or exogenous quinones. This signal has peaks at $g = 8$ and $g = 5.6$ and has been shown to be sensitive to the presence of herbicides and quinone analogues binding at the Q_B site. Detailed methods for sample treatment and EPR conditions required to observe the signal may be found in ref. 6.

3.3 The split phaeophytin signal

In order to permit the accumulation of reduced phaeophytin (Ph^-) in PSII, Q_A must be maintained in its reduced state. When PSII particles are illuminated at 200 K in the presence of 25–50 mM sodium dithionite to chemically reduce Q_A, oxidized electron donors can be re-reduced, but re-oxidation of Ph^- is prevented. Thus the Ph^- radical accumulates and can be detected by EPR. A 200 K bath is prepared in an unsilvered Dewar by the addition of ethanol to dry ice. The Ph^- signal in PSII is split by 3–5 mT due to an exchange interaction with Q_A- Fe^{2+}. The splitting is determined by the type of Q_A-Fe^{2+} present in the sample, either $g = 1.8$ or $g = 1.9$, the $g = 1.9$ form giving rise to a larger splitting. This can be observed in OGP core particles prepared at pH 6 and pH 7.5 (+ bicarbonate) as shown in *Figure 3*. The split signal is only observed at temperatures below 12 K and is best measured at 4–5 K. Above 12 K the normal Gaussian signal of the pheophytin is seen.

The Ph^- radical itself can be observed in the absence of Q_A- by illuminating PSII particles at room temperature in the presence of dithionite. This treatment results in the double reduction of Q_A. The Ph^- EPR signal which is centred around $g = 2.0035$ is also seen in the presence of dithionite in D1/D2/cyt b_{559} core complexes which completely lack Q_A and Q_B.

3.4 The primary electron donor

A spin polarized triplet signal ascribed to P680 is observed in PSII where Q_A is doubly reduced or absent as in reaction centre core particles. The *aeeaae* polarization pattern of the EPR spectrum is interpreted as originating from the radical pair recombination between $P680^+$ and Ph^-. To observe the spin polarized triplet signal in reaction centre core particles as prepared in section 2.4, transfer a sample containing 50 μg chlorophyll/ml to an EPR tube and freeze in liquid nitrogen. Record the spectrum at 4–5 K with the sample illuminated using a 150 W light source connected to the front of the cavity by a flexible light guide. Continuous illumination is required because the triplet decays with a half-life of approximately 1 msec at liquid helium temperatures. EPR conditions are given in *Figure 4*.

The absorption and emission peaks are arranged symmetrically around

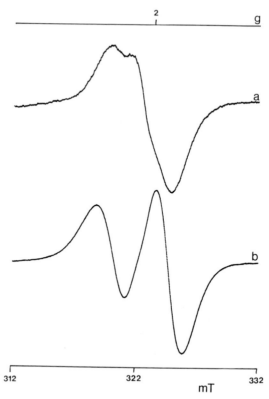

Figure 3. The split phaeophytin signals of OGP core particles following reduction with 25 mM sodium dithionite. (a) pH 6 OGP core particles, (b) pH 7.5 (+ bicarbonate) OGP core particles. Both spectra were obtained after illumination at 200 K and are light minus dark differences spectra. EPR conditions: temperature, 5 K; microwave power, 25 mW; modulation width, 0.4 mT.

$g = 2$, the positions of the peaks being given by the zero field splitting parameters D and E. These parameters give information on the magnetic environment and the interactions of the two unpaired electrons that give the triplet its spin $S = 1$. The zero field splitting parameters for reaction centre core particles as prepared in section 2.4 should be around $D = 0.0297$ cm^{-1} and $E = 0.0041$ cm^{-1}, consistent with a monomeric chlorophyll a.

The P680$^+$ radical is very difficult to observe since to allow its accumulation, electron donation by all paths as well as recombination of P680$^+$ with reduced electron acceptors must be prevented. However the $g = 2$ P680$^+$ signal can be observed in reaction centre core particles treated with silicomolybdate. Sample treatment and EPR conditions are described in ref. 7. Illumination of PSII centres at cryogenic temperatures also allows the oxidation of a monomeric chlorophyll (not P680) which gives rise to an

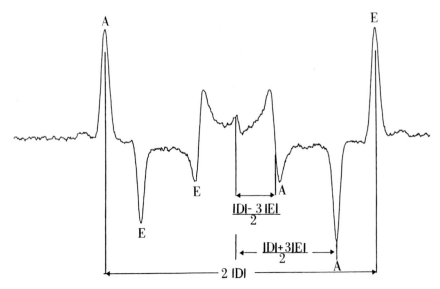

Figure 4. The spin polarized triplet signal of PSII reaction centre core complexes. The sample was illuminated in the cavity of the EPR spectrometer at 4.2 K. The spectrum shown is a light minus dark difference spectrum. EPR conditions: temperature, 4.2 K; microwave power, 50 μW; modulation width, 1.25 mT.

approximately 1 mT wide, $g = 2$ EPR signal superimposed on that of signal II.

3.5 Cytochrome b_{559}

Each PSII reaction centre contains a cytochrome with an absorption maximum near 559 nm. The role of cyt b_{559} remains unclear but it is thought to be involved in cyclic electron flow around PSII. Cyt b_{559} has a heterogeneous redox potential. In the light, oxygen evolving PSII particles exhibit mainly the high potential form (redox midpoint potential 375 mV), which becomes reduced on dark adaption. In further purified PSII preparations or in PSII particles from which the 17 and 23 kDa extrinsic polypeptides have been removed, lower potential forms are observed which are oxidized in the dark, as shown in *Figure 5*. To observe the high potential, oxidized cyt b_{559} signal, illuminate PSII particles at 77 K as described in section 3.2. At this temperature cyt b_{559} is the principal electron donor to P680$^+$. At about 10 K the oxidized cytochrome has EPR signals at around $g = 3$ (g_z) and $g = 2.2$ (g_y).

3.6 The S$_2$ state of the oxygen evolving complex

3.6.1 The multiline signals of the Mn complex

Under normal conditions P680$^+$ is reduced by electrons from the oxygen evolving complex, which probably contains four Mn atoms. The manganese

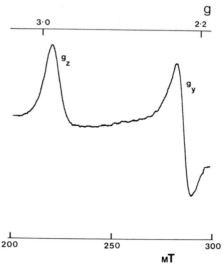

Figure 5. Low potential cytochrome b_{559} spectrum from reaction centre core particles. This signal was recorded in the dark. EPR conditions: temperature, 12 K; microwave power, 5 mW; modulation width, 1.6 mT.

complex is thought to provide the catalytic site for water oxidation and to be the site of charge accumulation: four turnovers of the reaction centre being required to produce molecular oxygen. During its cycle, the oxygen evolving complex passes through five different redox states, S_0 to S_4, O_2 being evolved at the S_3 to S_0 transition.

The S_2 state gives rise to a characteristic EPR signal known as the multiline signal (*Figure 6*). This signal is centred at $g = 2$, has > 18 lines, and is attributed to a mixed valence Mn cluster in an $S = \frac{1}{2}$ ground state. Following illumination of PSII particles, 75% of the centres will be in the S_1 state with the remaining 25% in the S_0 state. Dark adaption allows the slow oxidation of S_0 to S_1. A single saturating flash of light will generate the S_2 state which may be observed as the multiline signal if it is trapped by immediately freezing the sample to 77 K. The signal may also be generated by freezing under illumination in the presence of 100 µM DCMU or by illumination at 200 K. To do this, dark adapt PSII particles for 4 h on ice then freeze to 77 K. Prepare a dry ice–ethanol bath as above, in an unsilvered Dewar. Transfer the dark adapted sample to the dry ice–ethanol bath and illuminate each side of the sample for 5 min. Remove the sample and quickly wipe to remove ethanol from the surface of the tube prior to returning the sample to liquid nitrogen. EPR conditions are given in *Figure 6*.

Different types of multiline signal may be observed reflecting changes in the structure of the Mn complex. The inhibitory analogue of water, ammonia, binds to at least two sites in the oxygen evolving complex, one in competition

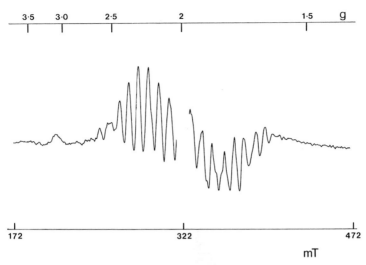

Figure 6. Multiline signal from PSII particles in the S_2 state. The sample was dark adapted for 4 h on ice, then frozen to 77 K, and the baseline recorded. The sample was then illuminated at 200 K for 5 min. The spectrum shown is the light minus dark difference spectrum. EPR conditions: temperature, 10 K; microwave power, 10 mW; modulation width, 1.25 mT.

with, and one independent of, chloride binding. Binding to the chloride site induces a stable $g = 4.1$ signal (see section 3.6.2). Ammonia binding to the second site modifies the multiline signal, reducing the line width of the signal and increasing its stability.

In the presence of a chelator, an altered dark stable form of the multiline signal is observed in PSII particles that have been depleted of calcium. Methods for calcium depletion may be found in ref. 8. The dark stable multiline (*Figure 7*) which is attributed to a modified S_2 state, is centred at $g = 1.98$ and has at least 26 lines with an average hyperfine spacing of 55 G and a half-life of several hours. If calcium or strontium are added to calcium depleted preparations the dark stable multiline is lost and illumination at 200 K results in the generation of the normal multiline signal (in the presence of calcium) or another modified multiline signal with reduced hyperfine splitting (in the presence of strontium).

3.6.2 The $g = 4.1$ signal

A second EPR signal is associated with the S_2 state of the Mn complex. This signal has a g value of 4.1 and is approximately 34 mT wide (peak to trough). This signal may be observed by illuminating PSII particles at temperatures below 200 K (140–160 K). This signal is attributed to a mixed valence Mn tetramer in an $S = > \frac{1}{2}$ state. Further discussion of the $g = 4.1$ signal may be found in refs 9 and 10.

Michael C. W. Evans et al.

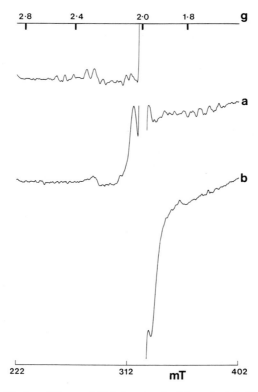

Figure 7. The modified multiline and 'S$_3$' signals of calcium depleted PSII. (a) Dark stable multiline signal from a sample of NaCl washed, calcium depleted PSII, illuminated at 277 K, and then dark adapted for 30 min. EPR conditions: temperature, 10 K; microwave power, 10 mW; modulation width, 1.25 mT. (b) 'S$_3$' split signal following freezing to 77 K under illumination. EPR conditions: temperature, 10 K; microwave power, 5 mW; modulation width, 1.25 mT.

3.7 The 'S$_3$' signal

Illumination of calcium depleted PSII particles at 273 K followed by freezing under illumination, results in a loss of the dark stable multiline signal and the generation of a 16.4 mT wide signal split around $g = 2$ (*Figure 7*). To obtain the maximum yield of this signal, an exogenous electron acceptor, PPBQ must be included in the EPR sample. EPR conditions are given in *Figure 7*. The detection of this signal is not limited to calcium depleted preparations and can also be observed in chloride depleted samples treated with fluoride, in ammonia treated PSII, and in the presence of acetate. For sample preparation and EPR conditions see refs 11 and 12.

The precise origin of the 16.4 mT 'S$_3$' signal is unclear. It has been assigned to S$_3$ since it occurs at one oxidation step higher than S$_2$, although it may not reflect the true S$_3$ state. The signal itself is thought to arise from an organic

198

radical magnetically interacting (exchange and/or dipolar coupling) with the Mn cluster. There have been two suggestions for the identity of the organic radical involved, histidine and tyrosine. Further discussion of the nature of the organic radical and origin of the S_3 signal may be found elsewhere (11, 12).

4. Preparation of photosystem I

Triton solubilized PSI can be conveniently prepared (*Protocol 6*) from the supernatant remaining after the isolation of PSII (section 2.2). PSI can be separated from the majority of the solubilized LHCPII complex by centrifugation, or prepared in more highly purified forms extensively depleted of the specific PSI light harvesting complexes by column chromatography (*Protocol 7*). We normally use hydroxylapatite chromatography (13), but ion exchange chromatography on DEAE cellulose is also successful. All of the steps during the purification procedure should be carried out at 4 °C in dim light.

Protocol 6. Preparation of Triton PSI particles by centrifugation

Regents

- 20% Triton X-100
- 1 M Tris–HCl pH 8.0
- 1 M K(PO$_4$) pH 6.8
- 2 M NaCl

Buffers used in this and *Protocol 7* are made by dilution from these stocks.

Method

1. Centrifuge the supernatant from *Protocol 2* at 145 000 *g* for 2 h to remove any remaining PSII.
2. Centrifuge the supernatant overnight (approx. 16 h) at 145 000 *g*.
3. Discard the green supernatant which contains LHCPII and free chlorophyll.
4. Rinse the surface of the pellet with 0.1 M Tris–HCl buffer pH 8.0 containing 0.1 M NaCl.
5. Resuspend the pellet in the same buffer using a paint brush. This is the photosystem I preparation.
6. Determine the chlorophyll *a/b* ratio and the P700/ chlorophyll ratio (*Protocols 9* and *10*).

Protocol 7. Purification of PSI by column chromatography

1. Adjust the PSI supernatant to 20 mM Tris–HCl pH 8.0 and 100 mM NaCl, then dialyse overnight in the dark at 4 °C against 20 mM Tris–HCl pH 8.0, 100 mM NaCl.

Protocol 7. *Continued*

2. Prepare a hydroxylapatite column containing 1 ml column volume of hydroxylapatite per millilitre of Triton extract to be treated. We normally treat about 200 ml per column containing 300–400 mg total chlorophyll. Pre-equilibrate the column with 20 mM Tris–HCl pH 8, 100 mM NaCl.

3. Apply the dialysed extract to the column.

4. Wash the column with 20 mM Tris–HCl pH 8, 100 mM NaCl, 0.5% Triton X-100 until the eluate runs clear. Then wash with three column volumes of 20 mM Tris–HCl, 100 mM NaCl, 0.1% Triton X–100 to remove excess detergent.[a]

5. Elute the column by slowly washing with 5–15 mM KH_2PO_4 pH 6.8, 100 mM NaCl, 0.1% Triton X-100. Collect fractions as the green band is eluted. Check the P700/chl ratio, pooling the best fractions.

6. Adjust the eluted sample to 20 mM Tris–HCl pH 8.0 and then dialyse for 12 h in the dark at 4°C against 20 mM Tris–HCl pH 8, 100 mM NaCl to remove the phosphate buffer.

7. Concentrate the dialysed fraction in an ultrafiltration cell over a 100 kDa membrane and store frozen. Alternatively the PSI can be concentrated by centrifugation at 160 000 *g* for 4–8 h.[b]

[a] PSI appears as a dark green band towards the top of the column.
[b] This avoids concentration of Triton micelles which may be a problem following concentration by ultrafiltration.

Protocol 8. Preparation of digitonin PSI particles (14)

Reagents

- 2% digitonin solution
- TEM buffer: 50 mM Tris–HCl, pH 8.0, 2 mM EDTA, 5 mM $MgCl_2$

Method

1. Prepare thylakoids as described in *Protocol 1.*

2. Add digitonin and TEM to the chloroplasts to a final concentration of 1.0 mg/ml chlorophyll and 0.5% digitonin. Stir and allow to digest for 30 min on ice in the dark.

3. Centrifuge at 15800 *g* for 30 min to pellet PSII. Discard the pellet.

4. Centrifuge the supernatant at 144000 *g* for 90 min. Resuspend the pellet containing the photosystem I in TEM buffer and store frozen until required.

4.1 Determination of chlorophyll concentration

The concentrations of chlorophyll *a* and *b* in a sample can be determined by the method described by Arnon (15) using the extinction coefficients for chlorophyll extracted in 80% (v/v) acetone (*Protocol 9*).

Protocol 9. Determining chlorophyll concentration

1. Dilute a suitable sample of the preparation (25–100 µl) in 10 ml of 80% (v/v) acetone.

2. Filter through Whatman No. 1 filter paper.

3. Measure the absorbance at 645, 652, and 663 nm.

4. Correct the absorbance for the dilution factor, then:

   (a) The contributions to absorbance by each chlorophyll at 663 and 645 nm is given by:

   $A_{663} = 82.04$ chl *a* $+ 9.27$ chl *b*
   $A_{645} = 16.75$ chl *a* $+ 45.6$ chl *b*

   where: A= absorbance, chl = chlorophyll concentration in mg / ml. When the above simultaneous equations are solved:

   chl *a* $= 0.0127\ A_{663} - 0.00259\ A_{645}$
   chl *b* $= 0.0229\ A_{645} - 0.00467\ A_{663}$

   (b) A value for the total chlorophyll, both *a* and *b*, can be calculated using the absorbance at 652 nm:

   chl $= 0.029\ A_{652}$.

The amount of P700 is determined (*Protocol 10*) from an ascorbate reduced minus ferricyanide oxidized difference spectrum using an extinction coefficient of 64 $mM^{-1}cm^{-1}$ at 697 nm and assuming a molecular weight of 893.5 Da for chlorophyll *a*. The extinction coefficient varies when different detergents are used (16).

Protocol 10. Determining concentration of P700

1. Dilute a sample of the preparation to 10 µg chl/ml.

2. Divide the sample between two 1 cm path length cuvettes.

3. Add 25 µl of 0.25 M sodium ascorbate to one cuvette, 25 µl 0.25 M potassium ferricyanide to the other. Record the difference spectrum between 680 and 720 nm. Determine the Δ_{OD} at the absorption peak.[a]

   [a] The peak may vary from 695 to 705 nm depending on the detergent used. The Δ_{OD} will be in the range of 0.001 to 0.05.

4.2 Measurement of NADP reduction

NADP reduction is a major physiological function of PSI. Measurement of NADP reduction is therefore the basic assay for intact PSI (*Protocol 11*).

Protocol 11. Measurement of NADP reduction

Reagents

- Chloroplasts or PSI preparation
- 100 mM sodium ascorbate
- 1 mM dichlorophenol indophenol (DCPIP)
- Spinach ferredoxin (prepared according to ref. 17)
- Spinach plastocyanin (prepared according to ref. 18)
- Spinach ferredoxin–NADP$^+$ oxidoreductase[a]
- 5 mM NADP

Method

1. At room temperature, prepare a reaction mixture containing:
   - 10 μg chl/ml
   - 10 mM sodium ascorbate
   - 100 μM dichlorophenol indophenol
   - 1.5 μM plastocyanin
   - 5 μM ferredoxin
   - 0.5 mM NADP$^+$
   - excess ferredoxin–NADP$^+$ reductase[a]
2. Measure the OD$_{340}$ of the sample.
3. Illuminate the sample with white light for 1 min. (Filter the light through a 3–5 cm water filter to prevent heating of the sample.)
4. Measure the OD$_{340}$ of the sample. The rate of NADP$^+$ reduction is given by:

$$\text{Rate in } \mu\text{mol/mg chl/h} = \frac{A_{340} \times 3600 \times 100}{5.85 \times [\text{chl}] \times \text{time (sec)}}$$

Rates of 5–700 μmol/mg chl/h should be observed.

[a] Partially purified preparations or commercial preparations may be used. The amount of the preparation required to saturate the reaction should be determined for each preparation.

4.3 Kinetic analysis of P700$^+$ re-reduction

Analysis of the re-reduction of P700$^+$ by back reaction from the bound electron acceptors is a useful technique for analysis of the electron acceptor complex and its modification by biochemical treatments. Reduction by back reaction from Fe–S$_{A/B}$ occurs with $t_{1/2} = 20$ msec, from Fe–S$_x$ with $t_{1/2} = 2$

msec, from A_1 with $t_{1/2} = 2$ μsec, and A_0 with $t_{1/2} = 20$ nsec. The experiments described here require measurements of the back reaction from the iron–sulfur centres. A spectrometer with time resolution of 100 μsec or better with actinic flash illumination of a few microseconds or less from a flash lamp or pulsed laster is required. The flash induced oxidation of P700 and its re-reduction is most easily followed at 810–820 nm. Use of this wavelength range minimize problems caused by absorption by the light harvesting chlorophylls and fluorescence from these chlorophylls. A relatively simple spectrometer using light emitting diodes as measuring beam source, diode detectors, amplifier, and a digital oscilloscope as signal averager is adequate for these measurements.

Protocol 12. Kinetic analysis of P700$^+$ re-reduction

Reagents

- PSI preparation
- 20 mM Tris–HCl, 0.1 M NaCl, pH 8.0
- 100 mM sodium ascorbate
- 1 mM dichlorophenol indophenol (DCPIP)
- 10 mM methyl viologen

Method

1. Prepare a solution of PSI containing 10 μg chl/ml in the Tris buffer.

2. Add 1 mM sodium ascorbate and 10 μM DCPIP to a sample and dark adapt for 2 min.

3. Determine the extent of the flash induced OD change at 820 nm and the $t_{1/2}$ for re-reduction.

4. Check that the OD change is proportional to the chlorophyll concentration.

5. Add 100 μM methyl viologen to a sample. The flash induced OD change should be lost as the P700 becomes oxidized when methyl viologen acts as electron acceptor.[a]

---

[a] If ascorbate and DCPIP are also added a slow re-reduction will be seen as electrons are donated from the dye.

4.4 Preparation of a photosystem I core complex

The Fe–$S_{A/B}$ protein and other low molecular weight peripheral polypeptides can be removed from PSI by treatment with 6.8 M urea. The procedure is based on that developed by Golbeck (19).

Protocol 13. Preparation of a PSI core complex

Equipment and reagents

- 9 M urea in 0.1 M Tris–HCl pH 8.0
- 20 mM Tris–HCl pH 8.0
- 2-mercaptoethanol
- Dithiothreitol
- Ultrafiltration membrane (100 kDa cut-off)

All solutions should be degassed and the procedure carried out under nitrogen or argon.

Method

1. Add 9 M urea dropwise with rapid stirring to a 1.5 mg chl/ml suspension of PSI particles.

2. Dilute the suspension with oxygen-free Tris–HCl pH 8.0 to 250 µg chl/ml and 6.8 M urea. Add 2-mercaptoethanol to 0.1%. Stir under argon at room temperature in the dark.

3. Monitor the removal of the Fe–$S_{A/B}$ protein by measuring the re-reduction kinetics of P700$^+$ by back reaction from the electron acceptors following flash excitation as in *Protocol 12*.

4. When the $t_{1/2}$ for P700$^+$ re-reduction changes from 20 ms and stabilizes at 1 msec,[a] stop the treatment by diluting the reaction mixture tenfold with oxygen-free 20 mM Tris–HCl pH 8.0, 5 mM dithiothreitol.

5. Remove urea by washing the preparation over an ultrafiltration membrane (100 kDa cut-off), or by centrifugation for 2 h at 150 000 *g*, and resuspension of the resulting pellet in 20 mM Tris–HCl pH 8.3, 0.1% 2-mercaptoethanol.

6. Adjust the preparation to approximately 1 mg chl/ml and store frozen until required.

[a] 25–30 min incubation for digitonin particles, shorter times for Triton preparations.

4.5 Removal of Fe–S_x from the PSI core particle

Incubation of PSI core particles in 5 mM potassium ferricyanide, 3 M urea, 20 mM Tris–HCl at 50 µg chl/ml for 30 min at room temperature results in the loss of the 1 msec P700$^+$/Fe–S_x back reaction kinetics. This indicates the oxidative destruction of Fe–S_x to the level of zero valence sulfur reported by Warren *et al.* (20).

Illumination of these samples at 200 K in the presence of dithionite at pH 10.0 results in the formation of identical EPR spectra of A_1 and A_0 as those observed in both the core particle and control PSI. The $g = 2.00$ EPR spectrum after 30 sec illumination is 0.89 mT wide, broadening to 1.4 mT with further 200 K illumination. This is indicative of a particle containing P700, A_0, and A_1, but depleted of iron–sulfur centres Fe–$S_{A/B}$ and Fe–S_x.

5. EPR analysis of photosystem I

P700 may be oxidized photochemically or chemically using potassium ferricyanide. The EPR properties of $P700^+$ are very sensitive to changes in the environment caused by detergents or the presence of oxidants or reductants. Quantitative measurements are therefore very difficult. The EPR signal of $P700^+$ is a Gaussian signal at $g = 2.0025$ $\Delta Hpp \approx 0.75$ mT, and can be observed at all temperatures from 4 K to 300 K. The intermediary electron carriers A_0 and A_1 are also organic radicals with narrow signals around $g = 2.0$. They have similar properties to $P700^+$ and are best observed at relatively high temperatures (above 77 K) and low microwave powers. The spectrometer should be set with a scan width of 5 mT centred at $g = 2.00$ and a modulation width of 0.1mT. The iron–sulfur centres, $Fe-S_{A/B/X}$ are transition metal centres with fast relaxation rates which can only be observed at low temperatures, below 25 K. $Fe-S_x$ has unusual properties and can only be seen easily between 6 and 10 K. The spectrometer should be set with a scan width of 100 mT centred at $g = 2.00$ with a modulation width of 1 mT or greater.

5.1 EPR measurement of $P700^+$

Add 20 µl of 10 mM K_3 Fe(CN)$_6$ to the 0.3 ml EPR sample. Incubate 1 min in the dark and freeze in liquid nitrogen. Alternatively, the sample may be illuminated for 1 min at room temperature without addition of oxidant. For room temperature spectroscopy it is best to illuminate the sample in the EPR spectrometer while recording the spectrum.

5.2 Iron–sulfur centre $Fe-S_A$ and $P700^+$

Carry out the following steps in the dark. Either incubate the sample for 1 h at room temperature then freeze in liquid nitrogen, or add 20 µl of 150 mM sodium ascorbate, incubate for 30 min, and then freeze. Transfer the sample to the spectrometer and record the spectra at temperatures below 25 K (around 15 K is best). Record the dark spectrum, illuminate the sample in the spectrometer. Record the light induced spectrum which includes $Fe-S_A$ ($g = 2.05$, 1.94, and 1.86) and $P700^+$ (*Figure 8*). Although these spectra are very similar in most organisms a few show some differences and the presence or absence of the peripheral proteins also affects the spectra (*Figure 9*).

5.3 $Fe-S_{A/B/X}$

(a) Add 20 µl of 2% (w/v) $Na_2S_2O_4$ to the EPR sample and illuminate for 2 min at room temperature. Turn the light off and freeze the sample in liquid nitrogen after 30 sec in the dark.

(b) Add 30 µl 2 M K-glycine pH 10.0, 20 µl 2% $Na_2S_2O_4$ and incubate for 30 min in the dark under N_2. Freeze in liquid nitrogen. Record the EPR spectrum at 15 K. The combined spectrum of $Fe-S_A$ and $Fe-S_B$ should

g

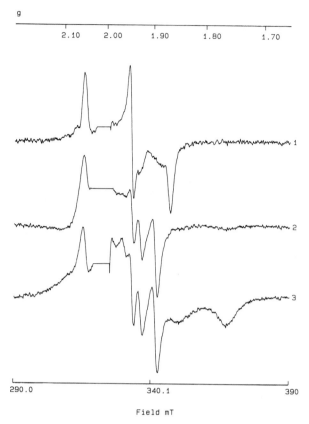

2.10   2.00   1.90   1.80   1.70

290.0   340.1   390

Field mT

Figure 8. EPR spectra of the iron–sulfur centres in spinach PSI. 1 Light − dark difference spectrum of a sample of spinach PSI prepared with Triton X–100. The sample was reduced with sodium ascorbate for 30 min in the dark, frozen in the dark, and then illuminated at 15 K in the spectrometer. The spectrum is that of the iron–sulfur centre Fe–S_A. EPR conditions: microwave power, 5 mW; modulation amplitude, 1 mT; temperature, 15 K; microwave frequency, 9.06 GHz. 2. A sample reduced with sodium dithionite illuminated for 1 min at room temperature and frozen with the light on. EPR conditions as for (1). The spectrum is that of the magnetically interacting iron–sulfur centres A and B. 3. The same sample as in (2) but with EPR conditions: microwave power, 25 mW; temperature, 8 K. The spectrum of iron–sulfur centre X is superimposed on that of A and B. The spectra are single scans recorded with a scan time of 2 min. The samples contained 0.5 mg chl/ml. The radicals due to P700$^+$, or A_1 and A_0 have been deleted.

be seen ($g = 2.05$, 1.92, 1.89) (*Figure 8*). Lower the temperature to 8 K, re-examine the spectrum around $g = 1.76$. Illuminate the sample in the spectrometer keeping the light on while recording the spectrum. Signals due to P700$^+$ ($g = 2.0025$) and Fe–S_{x-} ($g = 1.88$ and $g = 1.78$) will appear. The electron transfer from P700 to Fe–S_x is reversible at this temperature and the new signals will disappear when the light is turned off.

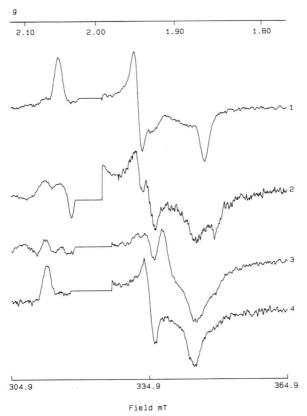

Figure 9. PSI is not identical in all organisms. EPR spectra of the iron–sulfur centres in PSI reduced with sodium ascorbate and illuminated at 15 K. 1. Spinach PSI as in *Figure 8*. 2. *C. reinhardtii* chloroplast membranes, showing reduction of Fe–S_A in some centres and Fe–S_B in others. 3. *C. reinhardtii* PSI isolated using Triton X–100. The spectrum resembles that of the isolated Fe–$S_{A/B}$ protein. 4. The *C. reinhardtii* sample shown (3) with the addition of the *psa*D gene product. The spectrum appears to reflect reduction only of Fe–S_B. (We are grateful to Professor D. Bryant for a gift of recombinant *psa*D gene product.) EPR conditions as in *Figure 8* (1).

The extent of reduction of Fe–$S_{A/B}$ observed in this experiment may vary between preparations. Triton preparations are more easily reduced than digitonin or untreated thylakoid membranes. Some irreversible oxidation of P700 coupled to reduction of Fe–S_B may be observed.

5.4 Fe–S_{x^-}, A_{0^-}, and A_{1^-}

(a) Add 20 μl 2% $Na_2S_2O_4$ to a sample prepared at either pH 8.0 or pH 10.0. Illuminate the sample for 1 min at room temperature and freeze in liquid nitrogen with the light on. These samples will have all of the iron–

sulfur centres in the reduced state with spectra as described above and will also have a $g = 2.00$ radical signal 1.2–1.6 mT wide which contains both A_{0-} and A_{1-}. The linewidth is affected by the detergent used.

(b) Prepare a sample as in section 5.3(a).

(c) Prepare a sample as in section 5.3(b).

(d) Prepare a second sample as in section 5.3(b), illuminate it for 1 min at room temperature, turn the light off and freeze after 10 sec in the dark.

These samples can be used to demonstrate the different components of the PSI electron acceptor complex using illumination at temperatures around 200 K to alter the redox state. Illumination may be done in a dry ice–ethanol bath or if a suitable temperature control system is available, in the EPR spectrometer.

With the spectrometer set to record the $Fe–S_x$ spectrum at 8 K, record the dark spectrum of a sample. Illuminate the sample at 205 K for 1 min, again record the spectrum. Note the increase in the extent of reduction of $Fe–S_x$ as indicated by the increase in the $g = 1.76$ signal. Repeat the illumination for suitable periods until there is no further increase.

With the spectrometer set to record the organic radical spectra around $g = 2.00$ (preferably at a temperature around 80 K) record the dark spectrum of a fresh sample. Illuminate the sample for 2 min at 205 K and record the spectrum again. A signal at $g = 2.0048$ with $\Delta Hpp = 0.95$ mT is observed, reflecting the reduction of the quinone acceptor A_1. Longer periods of 205 K illumination will result in the slow reduction of A_0.

Record the dark spectrum of a sample. Illuminate the sample at 230 K for 1 min and record the spectrum. Repeat the 230 K illumination recording the spectrum after increasing time intervals. Initially the spectrum of A_{1-} will appear, as the illumination proceeds the signal will broaden and change shape as A_{0-} appears. The shape and linewidth of the spectrum depends on the detergent used. In digitonin particles A_{0-} has $g = 2.0017$ and $\Delta Hpp \approx 1.7$ mT, in Triton particles the linewidth is narrower $\Delta Hpp = 1.2$ mT. At pH 8.0 the total $g = 2.00$ signal represents two spins per P700. At pH 10.0 the signal is larger, representing up to four spins per P700. At either pH a maximum signal size is reached which does not then increase with further periods of illumination. Similar spectra are observed in samples depleted of the iron–sulfur centres.

5.5 Double reduction of A_1

A_1 is a phylloquinone molecule. The spectra described above reflect reduction to the semiquinone state. Reduction to the quinol is not thought to occur during normal electron transfer processes. However the double reduction can be forced.

Prepare a sample as in section 5.3(b). Illuminate the sample for 90 min at

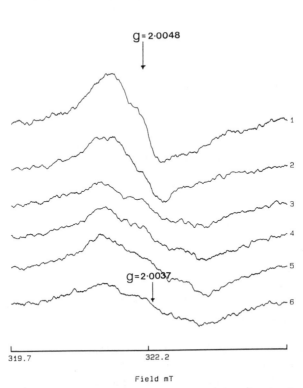

g = 2·0048

g = 2·0037

319.7   322.2

Field mT

Figure 10. The effect on the EPR spectra of A_{1-} and of A_{1-} plus A_{0-} in digitonin PSI particles of progressively longer illumination at 4°C pH 10.0 in the presence of sodium dithionite. Samples prepared as in *Protocol 8* were illuminated for 0, 2, 15, 30, 60, and 90 min (1–6, respectively). After freezing, the samples were subsequently illuminated for 2 min at 205 K. EPR conditions: temperature, 75 K; microwave power, 5 μW; modulation width, 0.2 mT.

273 K in an ice–water bath. At the end of this period freeze the sample in liquid nitrogen in the dark. Induce the $g = 2.00$ radical spectra by 205 K illumination as above. The spectrum observed has the characteristics of A_0 (*Figure 10*).

5.6 Photosystem I reconstitution assay

The Fe-$S_{A/B}$ depleted core complex (*Protocol 14*) described above can be reconstituted (*Protocol 15*) to give normal EPR spectra, kinetic behaviour, and NADP reduction using a partly purified Fe-$S_{A/B}$ protein preparation (21). Reconstitution of the 20 msec back reaction kinetics is routinely achieved with samples made with either digitonin or Triton preparations. Reconstruction of NADP reduction and the native EPR signals is achieved routinely with digitonin preparations. Triton preparations show variable results in these assays.

Protocol 14. Preparation of a partially purified Fe–S$_{A/B}$ holoprotein (22)

Reagents

- PSI particles prepared using *Protocol 6*
- Argon or oxygen-free nitrogen
- TEN buffer: 50 mM Tris–HCl, pH 8.0, 1 mM EDTA, 30 mM NaCl
- TM buffer: 50 mM Tris–HCl pH 8.1, 0.1% 2-mercaptoethanol

Method

1. Resuspend the PSI sample to 2 mg/ml chlorophyll in TEN buffer and remove dissolved O_2 by gassing for 3 h.
2. Add dithiothreitol to 5 mM.
3. Carefully layer an equal volume of ice-cold *n*-butanol over the sample and degas.
4. Vortex until the mixture appears homogeneous, then immediately centrifuge for 10 min at 30 000 *g*.
5. Isolate the aqueous phase containing the Fe–S$_{A/B}$ protein. Concentrate the sample and remove trace *n*-butanol by washing with oxygen-free TM buffer over an 5 kDa cut-off ultrafilter.

Protocol 15. Photosystem I reconstitution assay

Reagents

- PSI core preparation (*Protocol 13*)
- TMT buffer: 20 mM Tricene–HCl pH 7.8, 1% 2-mercaptoethanol, 0.07% Triton X-100
- Fe–S$_{A/B}$ holoprotein preparation (*Protocol 14*)

Method

1. Prepare the PSI sample at 12 μg chl/ml in TMT buffer.
2. Add a sample of the Fe–S$_{A/B}$ preparation. Incubate in the dark at 4°C for 10 min under argon.
3. Determine the P700$^+$ re-reduction kinetics (*Protocol 12*).
4. Repeat steps 1–3 with increasing amounts of Fe–S$_{A/B}$ preparation until the 20 msec decay time is fully restored.
5. Using a sample containing a threefold excess of Fe–S$_{A/B}$ protein as determined in step 4 determine the rate of NADP reduction.

6. Carry out a large scale reconstitution experiment using 300–500 μg of the depleted preparation. After incubation for 10 min wash the sample over an ultrafiltration membrane to remove unbound polypeptides, concentrate to 0.5–1 mg chl/ml. Prepare EPR samples reduced by the mercaptoethanol in the buffer and by sodium dithionite.

7. Obtain the EPR spectra of the samples to show the iron–sulfur centres, P700, and the light inducible changes at low temperature, comparing them with those of native preparations and the dithionite reduced Fe–S$_{A/B}$ preparation.

6. Studies of photosynthetic mutants of *Chlamydomonas reinhardtii*

6.1 Fluorescence screening of colonies

Examination of the stationary fluorescence yield is a useful method for screening a large number of colonies for photosynthetic mutants. These are detected by an increased or decreased stationary fluorescence (23). The basis of this method is that blue light ($350 < \lambda < 460$ nm) is applied to logarithmically growing colonies of cells. Fluorescence is observed by viewing the transmitted light through a red filter ($\lambda > 620$ nm). In the original method, stationary fluorescence is recorded photographically using Polaroid infra-red film. Since the film is now unavailable, we have avoided time-consuming photography by using a video camera linked to a monitor and thermal printer. This enables a rapid screening procedure with immediate identification of mutant colonies on the plates. You will need the following equipment:

- very bright white light source (> 120 W)
- blue filter ($350 < \lambda < 460$ nm)
- red filter ($\lambda > 620$ nm)
- closed circuit T.V. camera and monitor
- thermal paper printer
- image enhancer is a helpful addition if the light source is not sufficiently bright

An agar plate with up to 2000 colonies can be screened. The growth and maintenance of *C. reinhardtii* is fully described elsewhere (24).

The arrangement of the apparatus we have used is as follows: A 120 W tungsten-iodide light source is mounted behind a water filter and a blue filter such that 20 μE blue light is incident on the agar plate. The plate is viewed by a black and white closed circuit T.V. camera fitted with a red filter (Kodak Wratten 29). The colonies are viewed on a black and white monitor after enhancing the image using an Argus 10 image processor (Hamamatsu). Hard copies of the image are taken on a thermal printer.

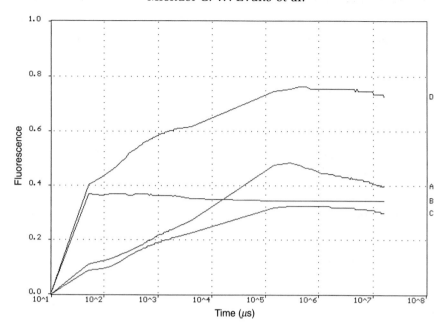

Figure 11. Fluorescence induction kinetics of photosynthetic mutants of *C. reinhardtii* (A) Wild-type; (B) mutant lacking PSII; (C) mutant lacking cytochrome b_6/f; (D) mutant lacking PSI. Fluorescence emission is expressed as arbitrary units (ordinate).

6.2 Fluorescence induction kinetics

With the commercial availability of equipment such as the Plant Efficiency Analyser (Hansatech), fluorescence induction kinetics can be monitored very simply in order to distinguish between high fluorescence mutants lacking PSII, the cytochrome b_6/f complex, or PSI (*Protocol 16*). The Plant Efficiency Analyser provides illumination from six light emitting diodes at 650 nm and fluorescence is measured above 730 nm. Methods for setting-up more traditional equipment are available (23). Specimen curves are shown in *Figure 11*. We have found it convenient to plot the fluorescence induction kinetics on a logarithmic scale. Although PSI and cytochrome b_6/f mutants have a similar induction curve, they can be distinguished by the lower relative fluorescence of the b_6/f mutants.

Protocol 16. Examination of fluorescence induction kinetics

Equipment

- Plant Efficiency Analyser with accompany-
  ing software
- IBM-compatible personal computer

Method

1. Grow a single colony of the strain of interest in the centre of a nutrient agar plate to mid-log phase (six days) in dim light (20 µE).

2. Place the plate in darkness for 20 min.

3. Place the plate over the clamped sensor head of the analyser. The colonies should be at the same distance from the light emitting diodes as optimized for a leaf disc by the manufacturers. This can be accomplished by adjusting the leaf clips.

4. Place black backing paper over the plate.

5. Take readings for at least 15 sec, and transfer the recorded data to the computer for analysis.

6.3 Preparation of *C. reinhardtii* samples for EPR analysis

For analysis of reaction centres I and II (the $P700^+$ and tyrosine Y_D. signals, respectively), only small amounts of material are required and packed cells are sufficient. For analysis of the Rieske iron–sulfur protein of the cytochrome b_6/f complex, thylakoid membranes should be prepared.

6.3.1 Analysis of reaction centres I and II

Grow the *C. reinhardtii* cells to mid-exponential phase (2×10^6 cells/ml). Pellet the cells by centrifugation at 3000 g for 10 min and resuspend the cells in 20 mM Tris–HCl, 5 mM EDTA pH 8.0. Repeat this twice, then resuspend the cells in the same buffer to 2–3 mg/ml chlorophyll. For rapid screening of mutant strains, EPR can be performed at room temperature using samples of 7 µl placed in heat-sealed capillary tubes. The baseline absorbance is substantial due to the absorbance of microwaves by water, but the light dependent signal for $P700^+$ is obvious. Clearer spectra, in particular for tyrosine Y_D·, can be obtained at 7 K with 0.3 ml samples frozen in the absence of illumination, as described in section 3. Examples of EPR spectra from mutant cells are given in *Figure 12*.

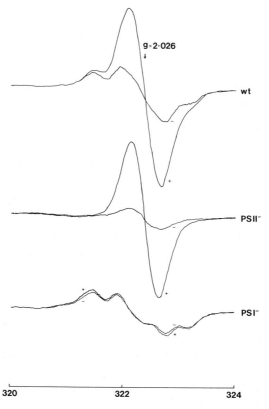

Figure 12. EPR spectra showing signals characteristic of PSI and PSII. Samples are whole cells of *C. reinhardtii*. In the dark the spectrum of PSII tyrosine radical $Y_D\cdot$ may be seen, whilst illumination at cryogenic temperatures generates the signal from P700$^+$ of PSI. wt: inner trace is the $Y_D\cdot$ spectrum of dark adapted cells, outer trace is the spectrum following illumination showing the generation of P700$^+$. PSII$^-$: as above, but showing the absence of $Y_D\cdot$. PSI$^-$: as above, but showing the absence of P700$^+$. EPR condition: microwave power, 10 mW; temperature, 8 K; modulation width, 0.2 mT.

6.3.2 Analysis of the cytochrome b_6/f complex

Protocol 17. Isolation of thylakoid membranes from *C. reinhardtii* (25)

Reagents

- HSM buffer: 10 mM Hepes–KOH pH 7.5, 0.35 M sucrose, 2 mM MgCl$_2$
- SHE buffer: 2.2 M sucrose, 5 mM Hepes–KOH pH 7.5, 5 mM EDTA
- Sucrose buffer: 0.5 M sucrose, 5 mM Hepes–KOH pH 7.5

Keep all solutions at 4°C.

Method

1. Culture 8 litres of the mutant in 10 litre carboys at 25 °C under dim light, stirring the culture and aerating by continuous bubbling of sterile air at 0.5 litre/min.

2. Harvest the cells at a density of $1-2 \times 10^6$ cells/ml by centrifugation at 3000 g for 5 min.

3. Resuspend the resulting pellets with a paint brush in 200 ml HSM.

4. Lyse the cells by passage through a French Press at 4000–6000 p.s.i.[a]

5. Check that lysis of the cells is complete by examination under a microscope. Repeat the lysis step if necessary.

6. Pellet the lysed cells by centrifugation at 48 000 g for 20 min.

7. Discard the supernatant and resuspend the pellet in SHE to a final sucrose concentration of 1.75 M (approximately 30 ml).

8. Place the resuspension in ultracentrifuge tubes, and carefully overlay with an equal volume of sucrose buffer. Centrifuge in a fixed angle rotor at 140 000 g for 2 h.

9. Isolate the thylakoid membranes which have banded at the interface between the two layers by removing first the sucrose layer and then the SHE layer using an aspirator. Dilute the membranes with sucrose buffer to give a chlorophyll concentration of 7–10 mg/ml.

[a] If the *C. reinhardtii* mutant is also cell wall deficient, the cells can be lysed by freezing and thawing the pellet prior to resuspension in HSM.

The EPR signal arising from the Rieske Fe–S centre of the cytochrome b_6/f complex is difficult to observe in membrane preparations due to the presence of other signals including that from the mitochondrial Rieske centre. However, the chloroplast Rieske can be identified by comparing spectra obtained in the presence of the electron acceptor PPBQ (phenyl-1,4-benzoquinone) or the specific cytochrome b_6/f inhibitor DBMIB (2,5-dibromo-3-methyl-6-isopropyl-1,4-benzoquinone). Membrane samples (0.3 ml) are prepared for EPR by mixing with either DBMIB or PPBQ (to 0.25 mM final concentration), and frozen under illumination. Typical spectra are shown in *Figure 13*.

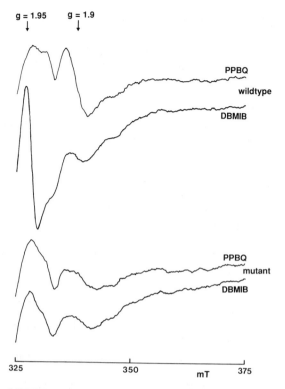

Figure 13. Use of DBMIB to identify the chloroplast Rieske centre by EPR. *C. reinhardtii* membranes were illuminated and then frozen to 77 K. In the presence of PPBQ a normal $g = 1.9$ signal is observed, but in the presence of the cyt b_6/f complex inhibitor, DBMIB the peak shifts to $g = 1.95$ (upper spectra). In a mutant lacking the complex the shift is absent (lower spectra). EPR conditions: microwave power, 5 mW; temperature, 15 K; modulation width, 1.6 mT.

References

1. Leegood, R. C. and Malkin, R. (1986) In *Photosynthesis energy transduction: a practical approach* (ed. M. F. Hipkins and N. R. Baker), pp. 9–26. IRL Press, Oxford.
2. Ford, R. C. and Evans, M. C. W. (1983). *FEBS Lett.*, **160**, 159.
3. Allen, J. F. and Holmes, N. G. (1986). In *Photosynthesis energy transduction: a practical approach* (ed. M. F. Hipkins and N. R. Baker), pp. 103–42. IRL Press, Oxford.
4. Dekker, J. P., Bowlby, N. R., and Yocum, C. F. (1989). *FEBS Lett.*, **254**, 150.
5. Nanba, O. and Satoh, K. (1987). *Proc. Natl. Acad. Sci. USA*, **84**, 109.
6. Bowden, S. J., Hallahan, B. J., Ruffle, S. V., Evans, M. C. W., and Nugent, J. H. A. (1991). *Biochim, Biophys. Acta*, **1060**, 89.
7. Nugent, J. H. A., Telfer, A., Demetriou, C., and Barber, J. (1990). *FEBS Lett.*, **255**, 53.

8. Hallahan, B. J., Nugent, J. H. A., Warden, J. T., and Evans, M. C. W. (1992). *Biochemistry*, **31** 4562.
9. Debus, R. J. (1992). *Biochim, Biophys. Acta*, **1102**, 269.
10. Kim, D. H., Britt, R. D., Klein, M. P., and Sauer, K. (1992). *Biochemistry*, **31**, 541.
11. Boussac, A., Zimmermann, J. L., Rutherford, A. W., and Lavergne, J. (1990). *Nature*, **347**, 303.
12. MacLachlan, D. J. and Nugent J. H. A. (1993). *Biochemistry*, **32**, 9772.
13. Williams-Smith, D. L., Heathcote, P., Sihra, C. K., and Evans, M. C. W. (1978). *Biochem. J.*, **170**, 365.
14. Boardman, N. K. (1971). In *Methods in enzymology* (ed. A. San Pietro), Vol. 23, pp. 268–76. Academic Press, New York.
15. Arnon, D. I. (1949). *Plant Physiol.*, **24**, 1.
16. Sonoike, K. and Katoh, S. (1990). *Plant Cell Physiol.*, **31**, 1079.
17. Katoh, S. (1973). In *Methods in enzymology* (ed. A. San Pietro), Vol. 23, pp. 408–13. Academic Press, New York.
18. Buchanan, B. B. and Arnon, D. I. (1973). In *Methods in enzymology* (ed. A. San Pietro), Vol. 23, pp. 413–40. Academic Press, New York.
19. Golbeck, J. H. (1987). *Biochim. Biophys. Acta*, **895**, 167.
20. Warren, P. V., Parrett, K. G., Warden, J. T., and Golbeck, J. H. (1990). *Biochemistry*, **29**, 6547.
21. Hanley, J. A., Kear, J., Bredenkamp, G., Li, G., Heathcote, P., and Evans, M. C. W. (1992). *Biochim. Biophys. Acta*, **1099**, 152.
22. Oh-oka, H., Takahashi, Y., Matsubara, H., and Itoh, S. (1988). *FEBS Lett.*, **234**, 291.
23. Bennoun, P. and Delepelaire, P. (1982). In *Methods in chloroplast molecular biology* (ed. M. Edelman, R. B. Hallick, and N.-H. Chua), pp. 25–38. Elsevier, Amsterdam.
24. Harris, E. H. (1989). *The Chlamydomonas sourcebook*. Academic Press, San Diego, CA.
25. Diner, B. A. and Wollman, F.-A. (1980). *Eur. J. Biochem.*, **110**, 521.

A1

Addresses of suppliers

Aldrich Chemical Co., The Old Brickyard, Gillingham LE17 4XN, UK; Milwaukee, WI, USA.

Amicon Corp., Upper Hill, Stonehouse, Gloucestershire, GL10 2BJ, UK.

Axon Instruments Inc., Foster City, CA, USA.

Beckman Instruments Inc., Spinco Division, 1050 Page Mill Road, Palo Alto, CA 94304, USA

Beckman Ltd., Turnpike Road, Cressex Industrial Estate, High Wycombe, HP12 3NR, UK.

Biorad Laboratories Ltd., Bio-Rad House, Maylands Avenue, Hemel Hempstead, Hertfordshire HP2 7TD, UK; 2000 Alfred Nobel Drive, Hercules, CA 94547, USA.

Bio-Logic Co., F-38640 Claix, France.

Dagan Corporation, Minneapolis, MN, USA.

Dow Corning, Midland, MI, USA.

Fluka Chemicals, Peakdale Road, Glossop, Derbyshire SK13 9YD, UK; 980 South Second Street, Ronkonkoma, NY 11779, USA.

Fluorochem Ltd., Wesley Street, Old Glossop, Derbyshire, SK13 9RY, UK.

Gilson Medical Electronics Inc., 3000 West Beltine Highway, PO Box 27, Middletown, WI 53362, USA.

Hamamutsu Phototronics UK Ltd., Lough Point, 2 Gladbeck Way, Windmill Hill, Enfield, Middlesex, EN2 7JA, UK.

Hansatech Instruments Ltd., Narborough Road, Pentney, King's Lynn, Norfolk, PE32 1JL, UK.

Hamilton Ltd., Kimpton Road, Sutton, Surrey SM3 9QP, UK.

Hilgenberg, PO Box 1161, D-34321 Malsfeld, Germany.

Hoefer Scientific Instruments, Box 77387, San Francisco, CA 94107, USA.

ICN Biomedicals Inc., 3300 Hyland Ave., Costa Mesa, CA 92626, USA.

Instrutech Corporation, Great Neck, NY, USA.

Intracel Ltd., Unit 4, Station Road, Shepreth, Royston, Herts SG8 6PZ, UK.

Kimble/Kontes Biotechnology, 1022 Spruce Street, PO Box 729, Vineland, NJ 08360, USA.

Lipex Biomembranes, Vancouver, Canada.

List-Electronic, D-6100 Darnstadt-Eberstadt, Germany; **Adams & List Ass. Ltd.,** Westbury, NY, USA.

Malvern Instruments, Spring Lane South, Malvern, Worcestershire, WR14 1AT, UK.

Merck Ltd., Poole, Dorset BH15 1TD, UK.

Millipore Corporation, 80 Ashby Road, PO Box 225, Bedford, MA 01730, USA.

Molecular Probes Inc., 4849 Pitchford Ave., Eugene, OR 97402–9144, USA.

Narishige Europe Ltd, London, UK; **Narishige USA Inc.,** Sea Cliff, NY, USA.

Nuclepore Corp., 63 Charlwood Drive., Oxshott, Surrey KT22 OHB, UK.

Pharmacia Biotech Ltd, 23 Grosvenor Road, St Albans, Herts AL1 3AW, UK.

Portex Ltd, Hythe, Kent CT21 6JL, UK; **Portex/Cincord Inc.,** Kit Street, PO Box 724, Keene, NH 03431, USA.

Radiometer Ltd., Manor Royal, Crawley, W. Sussex RH10 2PY, UK; Westlake, USA.

Rank Brothers, 56 High St., Bottisham, Cambridgeshire CB5 9DA, UK.

Seikagaku Kogyo Co. Ltd., Tokyo Yakugyo Bldg, 2–1–5 Nihonbashi-Honcho, Chu-ku, Tokyo 103, Japan.

Sigma Chemical Company, Fancy Road, Poole, Dorset BH17 7NH, UK; 3050 Spruce Street, St. Louis, MO 63103, USA.

Systech Instruments Ltd., Goodsons Industrial Mews, Wellington St., Thame, Oxon. OX9 3BX, UK.

G. Fredrick Smith Chemicals, Columbus, OH, USA.

Sutter Instruments, Novado, CA, USA.

William Freeman Ltd., Subaseal Works, Wakefield Road, Staincross ST5 6DH, UK.

Wilmad Glass Company, Route 40 & Oak Road, Buena, NJ 08310, USA.

Yellow Springs Instrument Company, Yellow Springs, OH 45387, USA.

Index

proteoliposomes *cont.*
  freeze-thaw 69–70
  fusion 71
  sonication 66–8
  protein orientation 74–5
  size/structure 71–5
  volume 72–3, 75
proton electrochemical potential, *see* protonmotive force
proton flux 1, 11, 33, 75–7
proton leak 6, 12, 120, 122–5
proton-linked transport 33
protonmotive force 22, 36, 39–62, 120–5
  see also membrane potential, pH gradient
proton permeability, *see* proton leak
proton stoichiometry, *see* H^+/e stoichiometry
pyranine 76, 77

quinones 98, 106, 188, 190–3, 207–9

radioisotope methods 22
radioisotope purity 32
rate constant 126
rate limiting step 126, 170
reaction centre, *see* photosynthetic reaction centres
reconstitution, *see* proteoliposomes
redox buffers 87
redox cells 96–102
redox couple 86–7, 93
redox electrode 98
redox interactions 108–9
redox mediators 87
redox potentials 8–10, 85–109
  measurement methods
    curve fitting and equations 102–9
    direct electrochemical 85–6
    EPR 96, 100–2
    equilibration with standard couple 86, 94–5
    spectroelectrochemical 86–94
  pH effect on 9, 93–4, 105–6
redox states 85
redox titrations 97, 100, 104–9
reducing agents 87, 89
regulation 111, 126–30
regulation analysis 126–7
resonance linewidth 164
respiratory chain 122–5
respiratory control ratio 6, 66, 67
response coefficient 113, 127
rhodamine 41
rubidium 41, 43

saturation transfer 163, 165–7, 169, 171, 178–9
sedimentation 28–30
site 1, *see* complex I
site 2, *see* cytochrome bc_1 complex
site 3, *see* cytochrome *c* oxidase
skeletal muscle 170
slip 8
sonication 66–8
soybean lipid 65
spectroelectrochemistry 86, 96
spin–lattice relaxation time 164, 166
spin–spin relaxation time 164
state 3 respiration 6
state 4 respiration 6
stoichiometry, *see* coupling ratio
submitochondrial particles 5, 9, 90
summation theorem 113
swelling 130, *see also* light scattering
sylgard 148, 152
symport 18, 22

thermodynamic coupling ratios 8–11
thylakoid preparation 185
top-down control analysis 115, 121–5
TPMP 41, 43, 50–61
TPMP/TPP electrode 43, 52–7, 122
TPP 41, 43, 50–61
transport, *see* membrane transport
transporter classes 35
trapped volume 72, 75

ubiquinone 8
uncoupler 120, 171

valinomycin 12, 40, 42, 77, 78, 129, 130
vesicles, *see* proteoliposomes
voltametry 86
volume, measurement
  in hepatocytes 58–9
  in mitochondria 44–7
  in proteoliposomes 72–3

xanthine oxidase 100, 103

yeast
  cell immobilisation 172–4
  growth and harvesting 138–9
  mitoplasts 138–41
  NMR 168, 172, 178–80
  phosphorylation potential 180
  plasma membranes 82
  P/O ratio 171